The Skeletal System

**Other titles in
Human Body Systems**

The Skeletal System

Evelyn Kelly

HUMAN BODY SYSTEMS
Michael Windelspecht, Series Editor

Greenwood Press
Westport, Connecticut • London

Library of Congress Cataloging-in-Publication Data

Kelly, Evelyn B.
    The skeletal system / Evelyn Kelly.
        p.   cm. — (Human body systems)
    Includes bibliographical references and index.
    ISBN 0–313–32521–9 (alk. paper)
    1. Human skeleton.   2. Human mechanics.   I. Title.   II. Human body systems.
    QM101.K44   2004
    612.7'5—dc22   2003067643

British Library Cataloguing in Publication Data is available.

Library of Congress Catalog Card Number: 2003067643
ISBN: 0–313–32521–9

First published in 2004

Greenwood Press, 88 Post Road West, Westport, CT 06881
An imprint of Greenwood Publishing Group, Inc.
www.greenwood.com

Printed in the United States of America

∞™

The paper used in this book complies with the
Permanent Paper Standard issued by the National
Information Standards Organization (Z39.48–1984).

10 9 8 7 6 5 4 3 2

Illustrations, unless otherwise credited, are by Sandy Windelspecht.

The *Human Body Systems* series is a reference, not a medical or diagnostic manual. No portion of this series is intended to supplement or substitute medical attention and advice. Readers are advised to consult a physician before making decisions related to their diagnosis or treatment.

To my husband, Charles L. Kelly,
who patiently assisted me in every way to make this
book happen.

To my four children, Sharlene, Kurt, Natalie, and Marsha,
who inspired me with their dedication to excellence.

To my four grandchildren, Aaron, Kyle, Kala, and Keenan,
who taught me love of life.

Contents

Color photos follow p. 126.

Series Foreword

Human Body Systems is a ten-volume series that explores the physiology, history, and diseases of the major organ systems of humans. An organ system is defined as a group of organs that physiologically function together to conduct an activity for the body. In this series we identify ten major functions. These are listed in Table F.1, along with the name of the organ system responsible for the activity. It is sometimes difficult to specifically define an organ system, because many of our organs have dual functions. For example, the liver interacts with both circulatory and digestive systems, the hypothalamus acts as a junction between the nervous and endocrine systems, and the pancreas has both digestive and endocrine secretions. This complex interaction of organs and tissues in the human body is still not completely understood.

This series is unique in that it provides a one-stop reference source for anyone with an interest in the human body. Whereas other references frequently cover one aspect of human biology, from anatomy and physiology to the prevention of diseases, this series takes a more holistic approach. Each volume not only includes a physiological description of how the system works from the cellular level upward, but also a historical summary of how research on the system has changed since the time of the ancients. This is an important aspect of the series, and one that is frequently overlooked in modern textbooks. In order to understand the successes and problems of modern medicine, it is important to recognize not only the achievements of the past but also the misunderstandings and challenges of the pioneers in medical research.

For example, a visit to any major educational institution reveals large

TABLE F.1. Organ Systems of the Human Body

Organ System	General Function	Examples
Circulatory	Movement of chemicals through the body	Heart
Digestive	Supply of nutrients to the body	Stomach, small intestine
Endocrine	Maintenance of internal environmental conditions	Thyroid
Lymphatic	Immune system, transport, return of fluids	Spleen
Muscular	Movement	Cardiac muscle, skeletal muscle
Nervous	Processing of incoming stimuli and coordination of activity	Brain, spinal cord
Reproductive	Production of offspring	Testes, ovaries
Respiratory	Gas exchange	Lungs
Skeletal	Support, storage of nutrients	Bones, ligaments
Urinary	Removal of waste products	Bladder, kidneys

lecture halls, where science instructors present material to the students on the anatomy and physiology of the human body. Sometimes these classes include laboratory sessions, but in the study of human biology, especially for students who are not bound for professional schools in medicine, the student's exposure to human biology typically centers on a two-dimensional graphic. Most educators accept this process as a necessary evil of the educational system, but few recognize that, in fact, the large lecture classroom is the product of a change in Egyptian religious beliefs before the start of the current era. During the decline of the Egyptian empires and the simultaneous rise of ancient Greek culture, the Egyptian religious organizations began to forbid the dissection of the human body. This had a twofold influence on medicine. First, the ending of human dissections meant that medical professionals required lectures from educators instead of participation in laboratory-based education, which the birth of the lecture hall. The second consequence would plague modern medicine for a thousand years. Stripped of their access to human cadavers, researchers studied other "lesser" animals and extrapolated their findings to humans. The practices of the ancient Greeks were passed on over the ages and became the basis for the study of modern medicine. These traditions continue to this day throughout the educational institutions of the world.

The history of human biology parallels the development of modern science. In the seventeenth century, William Harvey's study of blood circulation challenged the long-standing belief of the ancient Greeks that blood was produced in the liver and consumed in the tissues of the body. Harvey's pioneering experimental work had a strong influence on others, and within a century the legacy of the ancient Greeks had collapsed. In the eighteenth century, a group of chemists who focused on the chemical reactions of the human body, called the iatrochemists, began to apply chemical laws to human physiology. They were joined by the iatrophysicists, who believed that the human body must operate under the physical laws of the universe. This in turn led to the beginnings of organic chemistry and biochemistry in the nineteenth century, as scientists focused on identifying the building blocks of living cells and the chemical reactions that they utilize in their metabolism.

In the past century, especially in the last three decades, the rapid advances in technology and scientific discovery have tended to separate most sciences from the general public. Yet despite an ongoing trend to leave the majority of the physical sciences to the scientists, interest in the human biology has actually increased among the general population. This is primarily due to medical discoveries that increase not only lifespan but also healthspan, or the number of years that people live disease free. But another important aspect of this trend is the desire among the general public to be able to ask intelligent questions of their physicians and seek additional information on prescribed medications or procedures. In many cases this information serves as a system of checks and balances on the medical profession, ensuring that the patient is kept well informed and aware of the fundamentals regarding the procedure.

This is one of the most remarkable ages in the study of human biology. The recently announced completion of the Human Genome Project is an indication of how far biology has progressed. Barely fifty years ago, scientists were first discovering the structure of DNA. They now are in possession of an entire encyclopedia of human genetic information, and although they are not yet exactly sure what the content reveals, scarcely a week goes by without a researcher announcing a medical discovery that was made possible by the availability of the complete human genetic sequence. Coupled to this are the advances in the development of pharmaceuticals and treatments that were unheard of less than a decade ago.

But these benefits to society do not come without a cost. The terms stem cells, cloning, and gene therapy no longer belong to the realm of science fiction. They represent advances in the sciences that may hold the key to increased longevity. However, in many cases they also produce ethical and moral questions of society: Where do medical researchers obtain the embryonic stem cells for their work? Who will determine if humans can be cloned? What are the risks of transgenic organisms produced by gene therapy? These are just a few of the potential conflicts that face modern society. Only a well-educated general public can intelligently survey the pros and cons of an ethical or moral decision regarding medical science. Armed with information, concerned people can participate in the democratic process of informing their elected officials of their concerns. Science education is an important aspect of citizenship, and thus the need for series such as this to present information to the general public.

This volume covers the biology of the skeletal system. There are 206 bones in the human body, ranging in size from the massive femur of the leg to the miniature stapes in the ear. Although bones are crucial for support and movement, they also play other important roles in human physiology. The contractions of the heart muscle are made possible by calcium stored in the bones; the interior areas of long bones produce the cells of the circulatory system; and what we perceive as sound is a result of the action of small bones in the ear. We are all aware of the physiological consequences of damaging bones, including fractures and breaks, but few of us realize that as we age our bones progressively degenerate and that the influence of bone loss on health later in life is significant. Bones are dynamic, not static; they are constantly being reshaped according to the needs of the body. Osteoporosis, a disease characterized by calcium loss in the bones, affects both men and women later in life, resulting in fragile bones and immobility. The health of our bones is important; so is the need for a volume such as this one that is dedicated to the study of the human skeletal system.

The ten volumes of *Human Body Systems* are written by professional authors who specialize in the presentation of complex scientific ideas to the general public. Although any book on the human body must include the terminology and jargon of the profession, the authors of this series keep it to a minimum and strive to explain the concepts clearly and concisely. The series is ideal for the public libraries, as well as for secondary school and introductory college libraries. In addition, medical professionals or anyone with an interest in human biology would find this series a useful addition to their personal library.

Michael Windelspecht
Blowing Rock, North Carolina

Acknowledgments

I wish to express special thanks to those individuals who provided me support in researching and writing this book. I especially appreciate the series editor, Michael Windelspecht, and artist Sandy Windelspecht for their assistance, advice, and help in such a pleasant manner. Thanks also to the acquisitions editor, Debby Adams, for her encouragement and commitment to excellence.

I appreciate the staff members of the Central Florida Library, Ocala, Florida, who were most gracious in their assistance in locating texts as well as their endurance of my many trips for interlibrary loan.

I continue to be inspired by the many dedicated researchers and organizations who are committed to the health of the skeletal system.

Introduction

On top of a pile of bones, two skeletons jump with joy. Macabre bony figures gaze with amazement as another emerges from under a cloth. In the grisly woodcut *The Dance of Death*, Hartman Schedel captured the world of the Black Death, or plague, of the Middle Ages. Skeletons and bones expressed death's welcome respite from the horror (see illustration).

From ancient times, bones have fascinated human beings. In fact, bones are an integral part of ancient documents, literature, and art. The Hebrew Old Testament tells how God caused a deep sleep to come over Adam so that He could take one of Adam's ribs to make woman. Adam recognized, "This is now bone of my bone" (Genesis 1:28).

In the second century, the Greco-Roman physician Galen (129–216?) told medical students in his book *On Bones* the importance of studying osteology. However, Galen was able to study only the bones of animals and criminals who had not been buried. Human dissection was not permitted at that time.

Bones have often appeared as a symbol of danger or death. In the 1500s, pirates who plundered and terrorized ships and coastal towns hoisted flags bearing a skull and crossbones. The same symbol was at one time printed on labels of poisonous materials or used to mark hazardous places.

Throughout history, bones were associated with death because bone was considered to be dead. In fact, the word *skeleton* comes from the Greek word *skeletos*, meaning "dried up." It was not until the awakening of scientific investigation in the eighteenth century that an English surgeon, John Hunter (1728–1793), discovered that bone is living, dynamic, and changing.

The word for the study of bones, *osteology*, comes from two Greek words:

A woodcut by Hartman Schedel from the *Liber Chronicarum*, Nuremberg 1493, depicts the medieval concept of the Dance of Death. © National Library of Medicine.

osteo, meaning "bone," and *ology*, "study of." The branch of medicine that treats disorders and diseases of the bone system is *orthopedics*. The Frenchman Nicholas Andry (1658–1741) coined this term from two Greek words: *ortho*, meaning "straighten," and *paidos*, "child." Andry created a system of straightening the limbs using splints and exercise. Today osteology is the academic study of the anatomy and physiology of the bones; orthopedics is the branch of medicine dealing with diseases and disorders of the skeletal system.

Following a list of interesting facts about the skeletal system, this volume is organized into three sections. The first section (Chapters 1–6) presents basic information about the skeletal system. The second section (Chapters 7–13) focuses on the dedicated pioneers who explored the frontiers of the skeletal system from ancient times to the present. The third section (Chapters 14–18) presents selected medical information relating to diseases and disorders. The book concludes with a list of acronyms, a glossary of terms, a list of organizations and Web sites, a bibliography, and an index.

This book on the skeletal system is intended as a reference for students and interested readers, and an attempt has been made to describe medical and scientific concepts in common language. **Bold** type indicates the first use of key words that are listed in the glossary at the end of the book.

INTERESTING FACTS

- Eighty bones protect the vital organs of heart, lungs, spinal cord, and brain.

- Children with broken bones heal much faster than adults. A bone that requires three to five months for healing in an adult will mend in four to six weeks in a child.

- The spinal column consists of a series of 26 individual bones, or vertebrae.

- Motorcycle accidents account for one injury to the skeletal and muscular systems in every 7,000 hours of biking; horseback-riding accidents account for one injury in every 2,000 hours of riding—three and one-half times more than motorcycling.

- About 6.8 million people seek medical attention each year for injuries involving the skeletal system.

- Throughout the day the discs in the spine are squashed, making people shorter when they go to bed than when they wake up.

- Osteoclasts consume old and worn bone matter; osteoblasts manufacture new bone tissue. Both are important to good bone health.

- Sports medicine has been around since ancient times. To stay alive to fight, warriors had to keep in top physical condition.

- Nearly 40 million Americans—or 1 in 7—have arthritis, including 285,000 children.

- The average person will walk about 115,000 miles during a lifetime; that accounts for more than four jaunts around the equator on the feet.

- The feet have 52 bones, 35 joints, and 100 ligaments.

- A study of 1,122 students who carried backpacks to school found that 74.4 percent of the young people had severe to moderate back pain.

▶ All babies are born with flat feet. By age 12 or 13, the inner bones of the feet of most people have aligned to form arches.

▶ Osteoporosis, a bone-thinning condition, affects more than 25 million Americans, and 80 percent—or 20 million—are women. The disease causes more than 1.3 million fractures, including 500,000 spinal fractures, 250,000 hip fractures, and 240,000 wrist fractures.

▶ There are over 300 types of dwarfism (a condition resulting in stunted growth).

Functions of the Skeletal System

A formless jellyfish floats on the water, wafting back and forth with the wind and waves. Without bones, human beings would be just like this creature— gigantic, massive blobs that would move only by inching along the ground. Without bones, we would not be able to stand up straight, run, do handsprings, or even hear.

The skeleton provides shape and support to all the other body systems. In addition, it allows movement, protects body **tissues** and organs, stores important materials, produces valuable blood cells, and holds a record of our past development, diet, illnesses, and injuries.

PROVIDING SHAPE AND SUPPORT

In the human body, 206 bones are intricately arranged to keep the body upright. The skeleton is both rigid and flexible, enabling internal organs to defy the forces of gravity. The unique architectural plan makes the scaffold on which other body parts are hung and supported. One could even say the human being is a model of architecture and design. In fact, Marcus Vitruvius (first century BCE), an ancient Roman architect and engineer, advised his students who were designing symmetrical temples to study the human body because when a person's arms and legs are extended, he or she can touch a square with four corners and form a perfectly circular arc.

Body Plan

Compared with the nearest related primate, the gorilla, *Homo sapiens* is completely erect. Although the skull is at one end of the body, it functions

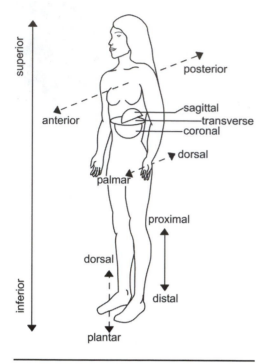

Figure 1.1. Planes.

as the center and main point of reference. As with any well-designed structure, form follows function. The classical anatomical form of *Homo sapiens* is standing erect, head straight, facing the observer, with arms at the side and palms facing forward. A midline perpendicular to the ground divides the body into left and right halves. This line represents the **sagittal plane**. A **transverse plane** runs parallel to the ground or floor.

To understand and read anatomy, one must be familiar with words that describe the positions of the body plan (see Figure 1.1). The following terms define the relationship of shape and support that create the framework of *Homo sapiens*

- **Superior** indicates toward the head; **inferior** indicates away from the head.
- **Anterior** refers to the front part of the body or body part; **posterior** refers to the back of the body or body part.
- **Medial** indicates toward the midline; **lateral** indicates to the side away from the middle.
- **Proximal** indicates closer to the torso; **distal** indicates farther away from the torso. For example, the hand is distal to the forearm because it is farther away from the torso than the arm.
- **Palmar** indicates the palm of the hand; dorsal indicates the back of the hand.
- **Plantar** indicates the sole of the foot.
- **Coronal plane** divides the body into front and back portions.

The terms may also relate to other **vertebrates** whose body plans are distinct from humans'.

The entire skeletal structure works together to manage the forces created by the upright position. When prehistoric animals walked on all fours, their back legs formed a right angle with the spine. The dog and cat reflect this general design. In the chimpanzee, which walks in a semi-upright position, the legs form less than a right angle. The upright position of the human causes the legs to form a straight line that runs through the backbone, the pelvis, and down each leg. When walking, as a leg on one side swings back, the arm on the same side swings forward. When taking a step, the opposite happens. The alternating movement keeps the body weight right in the center. (See photo in color section.)

When a person uses two feet, the stresses of walking, running, and jumping are calculated with precision. For example, when a person stands up, the hip takes half of the body weight, and the pull of stabilizing muscles can multiply this weight six times. When a person runs or walks, each hip alternates carrying the full body weight. Bones are extra strong at joints where large **compression** forces are generated.

The cranium and vertebrae support the brain and spinal cord. Whereas the cranium, or skull, shapes and supports delicate brain structures, the spine is the literal "backbone" of the human body. It forms a supporting rod for the head, arms, and legs. In other mammals the spine is a horizontal girder taking the weight off the chest and abdomen of animals that move on all fours. When the human spine is viewed from the side, an S-shaped curve is visible; this curve aids in balancing parts of the body over the legs and feet when standing. The chain-link arrangement of the vertebrae allows only a small amount of movement for each link; but when all the individual movements are added up, the spine is capable of making large, complex shapes and contortions. The S-shape helps bring the centers of gravity of the head, arms, chest, and abdomen above the legs. Thus the body as a whole is well balanced.

A Framework of Different Shapes

Each bone is designed to support, protect, and shape the body. For example, relative to body size, the human skull protects one of the biggest brains in the animal world. The features of different bones allow for efficient form and function. Bones may be classified by shape.

LONG BONES

Like tubular furniture, the tubular structure of long bones makes them both strong and light. The outer shell of bone is made of compact **cortical bone**. The spongy center has little **trabeculae**, or beams, that act as strengthening girders of **cancellous bone**. The inner casing of **trabecular bone** is arranged along lines of force with calcified fibers as they transmit the force into **tendons**, the fibrous bands that join muscles to bones.

Bones are thickest in the middle to support areas where forces are strongest. Weight for weight, bone structure is stronger than that of a solid rod. The lower extremities of bones are longer and stronger than the upper extremities. The purpose of the long bone of each leg, the femur, is to provide form and support and create an interconnected set of levers and linkages that allow movement.

SHORT BONES

Found in the wrist and ankle, these spongy, **cuboid bones** (shaped like cubes) are covered with a thin layer of compact bone. This arrangement permits shock absorption, movement, elasticity, and flexibility.

FLAT BONES

These bones in the ribs, the crest of the hip, the breastbone, and the shoulder blade are sandwiches of spongy bone between two layers of compact bone. They protect and provide attachment sites for muscles.

IRREGULAR BONES

Bones of the skull, face, vertebrae, and pelvis, as well as others that do not fit into another category, are referred to as irregular. Usually consisting of spongy bone with a thin compact bone exterior, they support weight and dissipate loads.

SESAMOID BONES

These are short bones embedded within a tendon or joint capsule. An example is the patella (kneecap), which allows the angle of insertion of a muscle to be altered.

The type of support provided by the bones reflects gender differences. Although men and women have the same number of bones, women's skeletons are lighter and smaller. Women's shoulders are narrow whereas their hips are broad and boat-shaped to accommodate a growing **fetus**. In men, proportions are reversed: broad shoulders and slim hips.

ALLOWING MOVEMENT

The skeletal system is an engineer's dream of a moving machine. "Mechanical science is of all the noblest and most useful, seeing that by means of this all animate bodies which have movement perform all these actions," said Leonardo da Vinci (1452–1519). Combining art and anatomy, he was one of the first to study how the interaction of the skeletal and muscular systems allows bodily movement.

Many parts of the skeletal system help the body to move. Tendons connect muscle to bone; **ligaments** connect bone to bone. Bones meet each other at **joints**. Muscles cause movement at joints by working in pairs. When one muscle contracts, the other extends. For example, the contraction of the biceps muscle in the front of the upper limb (humerus) and the relaxation of the triceps muscle behind this bone cause the elbow to bend. When the triceps contracts and the biceps relaxes, the arm straightens.

Rigidity of the bones allows for movement. The attached muscles at the joint permit freedom of movement in a variety of planes and in almost any direction.

Scientists are slowly learning more about how the human machine works. The field of bioengineering combines engineering principles with the anatomy of the living body to understand how movement occurs. For example, in a procedure called *gait analysis* subjects are fitted with diodes, or electronic sensors, that send out pulses of infrared rays of light as the subjects walk a prescribed route in the laboratory. Cameras sensitive to the emitted light follow the walkers' trails and record the positions relative to a fixed background. The cameras shoot as many as 315 frames per second. When this information is analyzed on a computer, a three-dimensional picture reveals if the person has problems with walking that arise from a skeletal deformity.

Bones as Levers

The science of physics is very important to human anatomy. Each time a person moves, the laws of physics come into play. The mechanics of movement integrates the laws of physics into biology. The skeleton is first and foremost a mechanical organ because one of its primary functions is to transmit forces from one part of the body to another. The tissues must bear loads without being damaged, and the skeleton must withstand very high forces because muscles can contract only a small percentage of their length. With the help of muscles the bones pull and push, creating the action of a lever. The term *fulcrum* refers to the point of support at which a lever turns in raising or moving something. This is the pivot point of the lever.

The underlying principle of physics that every machine reflects is the conservation of energy. Thus to simplify movements, levers save rather than spend forces. Moreover, the lever is one of the simplest machines. At the same time work (force) is exerted on one end of the lever, the other end moves the load. In the human body, muscles and bones function as one of three types of levers (see Figure 1.2).

CLASS 1 LEVER

In this class, the fulcrum is between the force and the load. A seesaw on a playground is an example. When force is exerted at one end, the load (the child) at the other end is lifted. In the human body, an example of a type 1 lever is the arm bending at the elbow. The elbow is the fulcrum. Contraction of the biceps muscle exerts force; the mass of the forearm bone is the resistance force. Another example is the head raising to look up. Muscles in back of the neck contract; bones of the face resist.

CLASS 2 LEVER

In this class, the load is between the effort force and the fulcrum. For example, exerting force on a long steel bar placed under an automobile frame

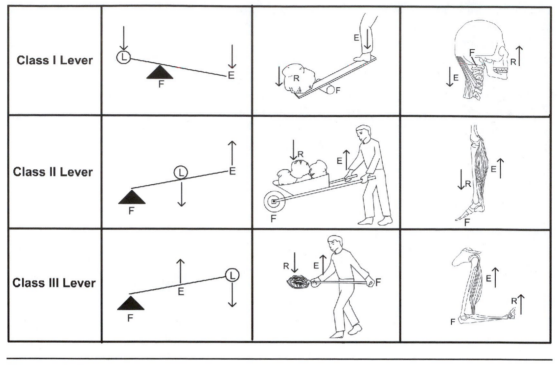

Class I Lever			
Class II Lever			
Class III Lever			

Figure 1.2. Levers.

lifts the car. Raising the foot to stand on tiptoe is an illustration in the human body. Gliding joints at the ankles, or tarsals, provide the fulcrum. Pushing down on the shin bone, or tibia, and the other leg bone, the fibula, provides resistance, while contracting the soleus muscle in the calf of the leg provides the effort force.

CLASS 3 LEVER

In this class, the fulcrum is at one end and the load is at the other end, with the effort force applied between. In the human body, the fulcrum is the elbow and the load is the hand. When the biceps move the hand, they function as a type 3 lever. This type of lever increases distance at the expense of force. When the biceps muscle moves a short distance, the hand moves a greater distance. The input and output forces are on the same side of the fulcrum and have the same direction.

Motion through Levers

How all these parts work together is demonstrated by a biomechanical marvel, the knee. The knee bone, or patella, hangs free from the center of the body and bends, glides, and rotates with the stresses that motion puts on it. Although four major ligaments and thirteen muscles support it, the knee is the most vulnerable of all joints because outside forces may displace these structures.

When a person takes a step the ligaments and muscles contract, relax, twist, and turn. Two long bones, the thigh and shin, sit between a piece of shock-resistant **cartilage** called *meniscus*. The crescent-shaped meniscus cushions and absorbs the shock of movement when walking or running.

To understand this function of movement, one must know several terms related to the action of bones and joints:

- **Flexion** involves bending or decreasing the angle at the joints. When the calf bends back toward the thigh, flexion occurs. When a bodybuilder flexes his muscles, he changes the angle of his bones at the joints.

- **Extension** is the opposite of flexion, with bones straightened to a 180 degree angle—a straight line.

- **Rotation** involves turning a body part on an axis. Just as the earth turns, the entire body may turn. However, a single body part (such as an arm or leg) cannot turn in a complete circle of 360 degrees because doing so would tear tissues such as blood vessels and nerves.

- **Abduction** involves drawing away from the midline of the body. Lifting up the arm at the shoulder joint moves it away from the body.

- **Adduction** involves moving toward the midline or trunk. Dropping the arm at the shoulder joint moves it back toward the body.

PROTECTING TISSUES AND ORGANS

The skeletal system protects the tissues and organs from the hard knocks of life. Although tremendous forces are constantly bombarding human bodies, bone is one of the strongest materials that nature has devised. One cubic inch can withstand loads of up to 20,000 pounds, which is about four times the strength of concrete.

The cranium, or skull, protects the soft delicate brain, which has the consistency of custard. The strong cranial case, which has an internal volume of about 2.5 pints, contains the organ that acts as "chief executive officer" of the body by controlling information from the outside world and responding to this information. Also protected are the organs of sight, hearing, smell, and taste. Eyes are set in sockets in the skull to stabilize the delicate structures. The delicate middle and inner ear not only are surrounded by bone but have three tiny bones that transmit sound. Through the bony recesses of the nose, air with vital oxygen passes on the way to the lungs. The jaws and teeth crush nourishing food on the way to the digestive system.

Attached to the cranium is the vertebral column, which consists of a series of small bones stacked on top of each other like a tower of spools. Large holes in the vertebrae line up to form a bony tunnel or canal. Nerves enter and leave the tunnel through gaps between neighboring vertebrae. Together, the vertebrae protect the delicate spinal cord, the important message cable between the brain and the other body parts. In front of the spinal cord, the rib cage and breastbone protect the heart, lungs, and part of the digestive system.

PRODUCING BLOOD CELLS

When a person gets a cut, a red fluid comes through the wound in the skin that contains red blood cells manufactured in the bones. Whenever there is a risk of infection or even a common cold, the body's defenders—white blood cells—are also made by the bones.

Indeed, an important function of the skeletal system is the production of blood cells. *Marrow* is a Latin word that means "middle"; hence, **bone marrow** is located in the middle, spongy part of bone. There are two types of marrow: Red marrow produces red blood cells, or **erythrocytes** and white blood cells or leukocytes; yellow marrow produces fat cells, or **adipoctyes**.

Red bone marrow is the site of **hematopoiesis**, or blood cell formation. In the adult human, this production takes place in bones such as the vertebrae, sternum, ribs, and pelvis and also at the ends of the upper arm (humerus) and the upper leg (femur). Red cells in the marrow make a substance called *heme*, which is an iron-containing nonprotein portion of **hemoglobin**, the red part of the erythrocyte. Hemoglobin contains iron and carries oxygen from the lungs to tissues. About 175 billion red cells per day are made and released according to the demands of the body.

The red marrow also produces white blood cells. About 70 billion per day of these important cells are needed for the body's defense. Also, about 175 million *platelets* per day are produced. These blood cells are important for clotting. According to demand, the system may increase production five- to ten-fold.

In newborns, red bone marrow fills in most marrow cavities. In older adults, much of the red marrow has been converted to yellow marrow. Long bones are filled with yellow marrow, which is mostly fat. In certain conditions like **anemia**, which occurs when there is a shortage of oxygen-carrying red blood cells, yellow marrow can be converted to red marrow for the manufacture of more red blood cells.

STORING MATERIALS

The bones are keepers of the minerals calcium and phosphorus. Deposited minerals account for about 50 percent of a bone's volume and 75 percent of its weight. In fact, 97 percent of the body's calcium is stored in bone.

Not only does the skeletal system store minerals, but it acts as a reservoir that maintains the **equilibrium**, or balance, of calcium and phosphorus in the bloodstream. Calcium **homeostasis** is very important for bodily functions. For example, too many calcium ions, or charged particles of calcium atoms, in the bloodstream can cause heart attacks; too few can cause respiratory problems. In the bones, these substances are available for rapid turnover when needed. For example, in pregnancy a growing fetus has a high demand for calcium. Storage in the bones makes available an extra supply. Likewise, after menopause, when menstrual activity ceases, changes in **hormones** may impair a woman's calcium and phosphorus levels, causing the minerals to leach out and leaving brittle, osteoporotic bones (see Chapter 5).

Many hormones play a role in this storage system. These include **estrogen**, testosterone, thyroid hormone, adrenal gland hormones, insulin, and growth hormone.

GIVING CLUES TO THE PAST

The skeleton's durability is staggering. Even after death it has many uses. Archaeologists, anthropologists, and forensic scientists can glean valuable information from bones as to what happened in the past. For example, fossils (bones that have turned to stone) offer a broad outline of how the human face evolved. Fossils of *Australopithecus*, the Southern Ape that existed 2 to 3 million years ago, can be compared to those of *Homo sapiens neanderthalensis*, or Neanderthal Man, of 100,000 years ago. Both of these can be compared to the first *Homo sapiens,* or Thinking Man, who lived 40,000 years ago. Gradually the face became flatter, the teeth smaller, the chin less protruding, and the forehead more domed to house a larger brain. (See "Dr. Donald Johanson and Lucy.")

Bones also tell the story of humans' adaptive mechanisms to the environment over time. Bones often survive the process of decay and are important in the following ways:

- They provide evidence of fossil man.
- They form the basis of racial classification in prehistory.
- They give information about culture and people's world view.
- They are major sources of information about ancient disease and causes of death.
- They help solve forensic crimes by providing evidence to detectives.

Different cultures have used bones in various ways. Some African tribes made ceremonial drums from the skulls of enemy warriors. The holy men

Dr. Donald Johanson and Lucy

In 1974 while exploring at Hadar, Ethiopia, Dr. Donald Johanson (1943), a contemporary American paleoanthropologist (who studies fossils of ancient man), made a lucky find. He discovered a complete skeleton that he named *Lucy* (after the Beatles' song "Lucy in the Sky with Diamonds"). About three million years ago Lucy apparently walked near a stream, fell into the water, and drowned. Her body sank into the mud, her flesh slowly decomposed, and the calcium in her bones was replaced by minerals in the water. Over eons of time the water turned into desert. Johanson was at the right place when rainwater washed away the dirt from the dried-up lake bed and exposed Lucy's body.

Johanson calculated that Lucy's brain was about one-third the size of a modern human brain, but she had an important human characteristic: Her knee could lock so that she could stand up straight. Apes and other primates do not have locking knees, the structure that allows standing for a long period of time. Lucy was evidently a meat-eater, differing from other primitive humans who were vegetarians. As their brains grew larger their bodies had a need for **protein**, so they became carnivores. Johanson surmised that Lucy was a scavenger who fed on the carcasses of dead animals. Lucy represents one of the oldest hominids in the fossil record. *Hominid* is the name of the family of humankind and its ancestors.

The next year at the same site, Johanson found the fossil remains of some thirteen individuals that he believed to be human ancestors living in groups. The scientific name of Lucy and the community groups is *Australopithecus afarensis*.

Johanson has spent the last twenty-five years exploring Africa, Ethiopia, Tanzania, Yemen, Saudi Arabia, Egypt, and Jordan looking for bone remains that tell more of the history of humans. In 1986, in the Olduvai Gorge in Tanzania, he discovered a 1.8 million-year-old partial skeleton of *Homo habilis*, the first toolmaker.

of Tibet used ceremonial cups made from the top of the human skull as a respectful symbol of committing the mind of one individual to another.

The human skeletal system is a marvel of mechanical and architectural design. It is intricately structured to keep the body straight and upright but flexible enough to permit great freedom of movement. It is strong enough to protect vital organs while also producing blood cells and storing and regulating minerals. As a source of data, bones have made possible the study of the evolution, history, and culture of *Homo sapiens*.

Bones of the Central Skeleton: The Axial System

Osteology—the study of bones—is important to many scientific disciplines, including archaeology, physical anthropology, geology, paleontology, anatomy, medicine, and forensics. In fact, knowledge of the framework of the skeleton is essential not only for scientists but also for lay people. How the body's framework functions is essential to developing good health habits and care of the skeletal system.

The sturdy scaffold of the human body is made up of 206 bones. Softer tissues and organs are attached to this structure. Actually, the skeleton, directly or indirectly, supports or connects to all body parts. Bones are grouped into two categories: the **axial skeleton** and the **appendicular skeleton**. This chapter discusses the axial skeleton; bones of the skull, vertebral column, ribs, and breastbone. Chapter 3 features the appendicular skeleton: bones attached as appendages to the axial skeleton.

SPECIAL BONE STRUCTURES

Some structures are important in understanding both the axial and the appendicular skeletons. These structures include **sutures** and **processes**.

Sutures

During birth, a baby's head is squeezed as it passes through the birth canal. In order for the baby to be born safely, its skull is not solid bone. A **membrane** covers the areas where the skull bones have not yet grown together. The structures known as *fontanels*, or soft spots, allow the bones of

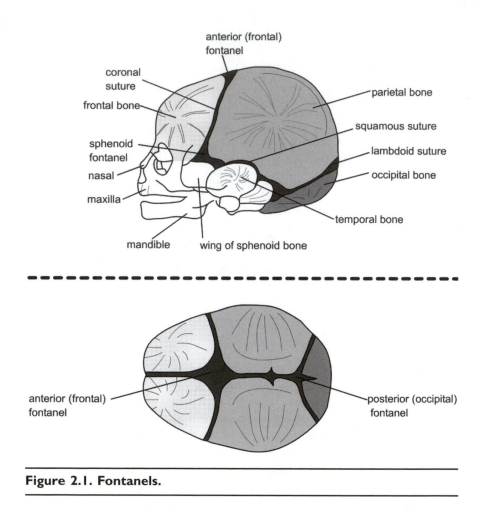

Figure 2.1. Fontanels.

the skull to mold, slide, and overlap to minimize danger to the delicate brain during birth (see Figure 2.1). These soft areas disappear completely by age 2, closed by structures called *sutures* (see photo).

Sutures could be referred to as bone zippers as they occur where the bones of the skull come together along serrated and interlocking joints. The areas appear as irregular gaps before the age of 17 but grow together as the person gets older. By age 30 or 40, the sutures gradually fade away. Looking at these sutures is one way of telling the approximate age of a skull.

As the sutures turn into bone, the process (the outgrowth of bone) begins inside the skull and then knits together toward the outside. Most sutures are named from the bones that grow together to form them. For example, the *ethmoidofrontal* suture joins the *ethmoid* and *frontal* bones. Other special names for sutures follow (see Figure 2.2):

- *Coronal:* between the frontal and parietal bones

- *Sagittal:* between the two parietal bones

- *Basilar:* between the occipital and the sphenoid bone

- *Squamosal:* between temporal and parietal bones

Processes

A process is a projection or outgrowth of bone or tissue. On any given bone, depressions and holes may also be present. These structures provide for the attach-

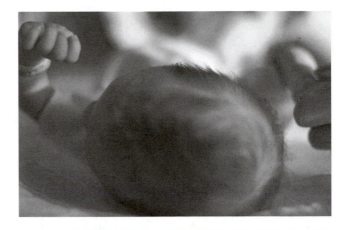

The fontanel, or soft spot, of a newborn baby. © Rubberball Productions.

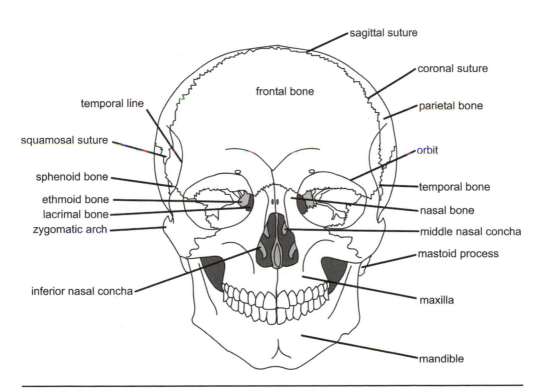

Figure 2.2. Skull, anterior.

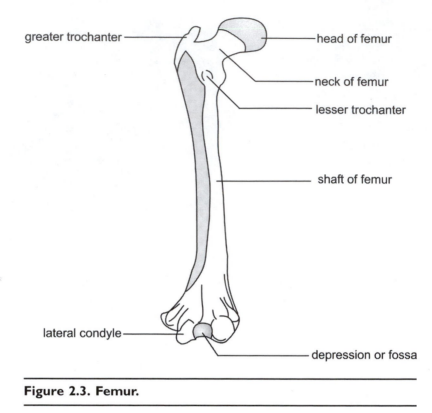

greater trochanter ——————————————— head of femur

——— neck of femur

——— lesser trochanter

——— shaft of femur

lateral condyle ——————————

—— depression or fossa

Figure 2.3. Femur.

ment of muscles, help form joints, and act as passageways for blood vessels and nerves (see Figure 2.3).

Bone processes include the following structures:

- *Head*—a large round part that joins with another bone; the term **articulation** refers to the joining of one bone to another
- *Shaft*—the principal part of a long bone
- *Neck*—the narrow part of a bone between the head and shaft
- *Spine*—a sharp slender process, as seen on the back of the shoulder blade (scapula)
- *Condyle*—a rounded, knuckle-like process located where one bone articulates with another
- *Crest*—a very narrow ridge of bone
- *Trochanter*—a large projection for the attachment of muscles
- *Depression*, or **fossa**—a shallow hole in the surface of the bone

THE AXIAL SKELETON

The axial skeleton—the central supporting portion of the body—is composed of skull, vertebral column, ribs, and breastbone. The term *axial* is derived from the word *axis*, a real or imaginary straight line that runs through the center of a body. *Axis* may also refer to a structure around which other objects rotate. The axial skeleton has eighty bones.

The Skull

The skull (see illustration) is divided into the cranium and facial bones. Oval in shape and wider behind than in front, the skull rests on the top of the vertebral column. The human skull starts life as a "jigsaw puzzle" of about thirty pieces held in cartilage and membranes. During embryonic development, these pieces gradually grow together to form a solid case. There are six fontanels at birth; the pulsing of the baby's blood system can be seen in the uppermost fontanel. The cranium, which lodges and protects the brain, consists of eight bones. The names of these eight bones, along with their common names or locations, follow (see Figure 2.4):

- *Frontal*—the forehead

- *Occipital*—lower back of the skull

- *Sphenoid*—large bone between the occipital and ethmoid in front and temporal bones at the side

- *Ethmoid*—inner part of the eye socket and back of the nose

- *Two sets of parietal*—top and side of the skull

- *Two sets of temporal*—temple areas on the side of the skull

These flattened or irregular bones do not move—with one exception, the mandible (jaw). They are joined at points called *sutures.*

FRONTAL BONE

The frontal bone forms the forehead and the upper portion of each eye socket, or orbit. It has two parts: a vertical portion

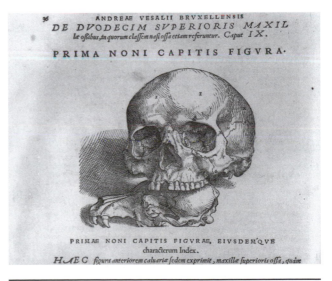

Front view of the skull with the jawbone removed, contrasted to an animal skull. From Vesalius, *De Humanis Corporis Fabrica,* 1543. © National Library of Medicine.

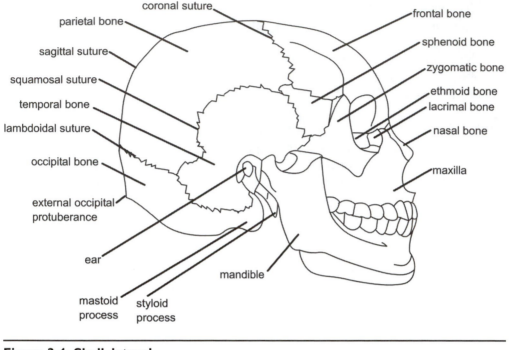

coronal suture
frontal bone
parietal bone
sphenoid bone
sagittal suture
zygomatic bone
squamosal suture
ethmoid bone
temporal bone
lacrimal bone
lambdoidal suture
nasal bone
occipital bone
maxilla
external occipital
protuberance
ear
mandible
mastoid process   styloid process

Figure 2.4. Skull, lateral.

called the *squama* in the forehead region, and a horizontal portion that forms the roofs of the orbital and nasal cavities.

The outer surface of the squama is convex and usually shows the remains of the frontal, or metopic, suture that divides the bone in two. On each side of this suture, located about 3 centimeters above the eye socket, is a rounded elevated area called the *frontal eminence*. These protrusions vary in size and are usually larger in men than in women. Sometimes they are referred to as *supraorbital ridges*. The frontal bone articulates with twelve other bones.

OCCIPITAL BONE

The occipital bone forms the posterior, or back, surface of the skull. It is shaped like a trapezoid—a four-sided figure with two parallel sides and two nonparallel sides. The structure seems to curve in on itself. A large oval hole, or aperture, called the *foramen magnum* pierces the bone. This hole, the largest foramen, allows the spinal cord to pass through the bone.

The external surface of the occipital bone curves out in a convex area, producing a protrusion on either side to which ligaments are attached. The lateral, or side, parts of the occipital bone rest at the side of the foramen magnum.

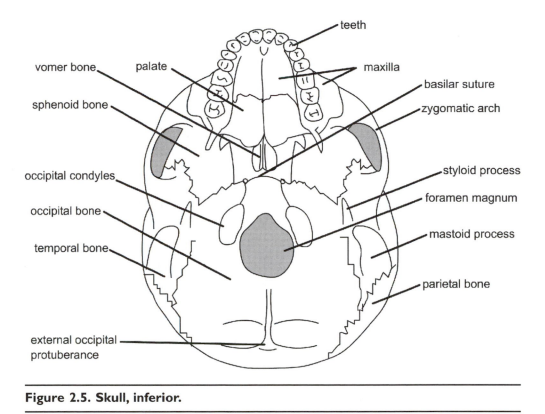

Figure 2.5. Skull, inferior.

SPHENOID BONE

One of the most difficult bones to describe is the sphenoid. A number of features and projections enable it to be viewed from various points. A single bone, it runs through the **midsagittal plane** (see Figure 2.5) and connects the cranium to the facial bones.

The sphenoid bone is a hollow body that contains the sphenoid sinus and three pairs of projections. On the inside of the sphenoid is a small saddle-shaped shelf where the **pituitary gland** rests. One section of the sphenoid bone called the smaller lesser wings has a hole that allows the optic, or second cranial, nerve to pass through. Other processes run along the back portion of the nasal passages toward the palate, or roof of the mouth. Muscles run from these attachments to the internal, or medial, surface of the mandible, or jawbone. These muscles provide the grinding motion of chewing.

ETHMOID BONE

The ethmoid bone is a light and somewhat spongy bone that is cubical. A single bone, it runs through the midsagittal plane that connects the cranium to the facial skeleton. Unlike the sphenoid, one cannot see it from various views.

The bone has paired projections called the *Crista Galli*, or *Cock's Comb*. Several plates make up this single bone, which cradles the nerve of smell, separates the nasal passage, and forms part of the eye socket.

ETHMOID NOTCH

The ethmoid notch separates two orbital plates. In front of the notch are the openings to the frontal air sinuses. A *sinus* is a cavity within a bone or organ. In the interior of the cranium, these cavities are vein channels that move blood from the brain. External sinuses—such as those found in the frontal, sphenoidal, ethmoid, and maxillary (upper jaw) areas—are hollow spaces in the bone. They are connected to the nasal cavities that contain air.

PARIETAL BONES

Between the frontal and occipital bones are two parietal bones. As the two bones unite, they form the sides and roof of the cranium. Each bone is roughly quadrilateral and has two surfaces, four borders, and four angles.

TEMPORAL BONES

The temporal bones are paired cranial bones situated at the side and base of the skull. They are located below the parietal bones and form part of the sides of the base of the cranium. Each temporal bone contains the middle and upper portion of the hearing mechanism. One area of the temporal bone forms the *mastoid process*, a conical projection. The mastoid can be felt by placing the hand behind the ear. The mastoid process is larger in men than in women. Part of the temporal bone extends out to form the *zygomatic arch* (see next section) of the facial bones.

Facial Bones

All bones of the face are paired except the *vomer, mandible*, and *hyoid*. The paired bones are the following:

- *Zygomatic*
- *Maxillae*
- *Nasal*
- *Lacrimal*
- *Palatines*

ZYGOMATIC

The cheekbone is made up of two bones; the zygomatic, and a fingerlike projection from the temporal bone. This joining is called the *zygomatic arch*. Also called *malar* or *jugal*, each zygomatic bone articulates (joins) with surrounding bones, one on each side.

MAXILLAE

The maxillae, or upper jawbone, are paired facial bones that join to form the hard palate in the roof of the mouth. They also contain the upper teeth.

NASAL

Some facial features are composed of both bone and cartilage. The nasal bones are small rectangular bones that form the upper part of the bridge of the nose. Cartilage forms the lower part of the nasal frame. Cartilage deteriorates after death. This is why one never sees a skeleton with a nose.

LACRIMAL

The lacrimal bones are located behind and lateral to the nasal bones. These small and fragile bones help form the eye orbit and part of the nasal passage. They also contain the fossae, or holes, housing the lacrimal duct that connects the medial corner of the eye to the nasal passage. This duct enables tears from the eye to enter the nasal passage.

OSSICLES

The human body's tiniest bones are in the ear. Three little bones called *ossicles* are located in each middle ear. The bones are named for their appearance:

- The *malleus*, or hammer, is about 0.32 inch long.
- The *incus*, or anvil, looks like a small version of the metal table used by blacksmiths to hammer iron tools.
- The *stapes*, or stirrups, appear as a saddle about 1.2 inches long.

The mallet-shaped handle of the malleus, or hammer, attaches to the inner surface of the eardrum. The head of the hammer fits into a tiny socket at the base of the anvil. Small ligaments between these two bones bind them firmly together. A long process of the anvil joins with the head of the stirrup. Hearing occurs when sound waves strike the eardrum and set the ossicles in motion. Moving through the bones, the waves result in a rocking motion of the stirrups that oscillates against the membrane covering the opening (called the *oval window*) in the inner wall of the middle ear. At birth, the ossicles are completely developed. They do not change in size as a person grows.

VOMER

The vomer is a thin, flat bone that looks like a plowshare. Joining with the ethmoid bone, it becomes the partition between the two nasal cavities, or septum. The lateral walls of the nasal cavity have two scroll-shaped bones called *inferior nasal conchae*. These thin, porous paired bones are elongated and curl upon themselves. Attached to the wall of the nasal cavity, the bones increase the amount of mucus membrane and olfactory nerve endings that contribute to the sense of smell.

MANDIBLE, OR JAWBONE

The largest and strongest bone of the face is the mandible. This horse-shoe-shaped bone holds the lower teeth. Its parts include a curved, horizontal body and two perpendicular portions, the *rami*, that join the ends at right angles.

In adults, portions of the base are of equal depth and the rami are almost vertical, measuring from 110 to 120 degrees. However, in old age the bone is greatly reduced in size. With the loss of teeth some bone is absorbed and each ramus becomes oblique, with an angle that measures about 140 degrees.

HYOID

The U-shaped hyoid bone is unique in that it does not attach to any other bone. Located in the neck above the larynx, or voice box, it serves as the attachment for the muscles of the tongue.

The Vertebral Column, or Spine

The vertebral column is the backbone of the body, forming a supporting rod for the head, arms, and legs. The vertebral column is made up of a series of twenty-six bones called *vertebrae* (singular, *vertebra*). Cartilage and ligaments link the vertebrae together, allowing flexibility and giving support to the trunk. The column protects the spinal cord.

Vertebrae are grouped according to the region they occupy (see Figure 2.6):

- *Cervical*, or neck—seven
- *Thoracic*, or chest—twelve
- *Lumbar*—five
- *Sacral*—five fused bones between the hip bones
- *Coccygeal*—tailbone region

In the embryo, thirty-three separate vertebrae exist. Before birth, the five sacral vertebrae and four coccygeal vertebrae fuse, forming a single bone at birth.

Seen from the side, the spinal column does not form a straight line. Several curves correspond to the sacral curves that appear before birth, and others develop later. The cervical curve in the neck appears when an infant begins to sit up and hold its head erect. Another forms in the lumbar region when the baby begins to walk. Changes in the curvature of the column result in shifts in the body's center of gravity. The average length of the spinal column in men is 28 inches (71 centimeters); in women, 24 inches (61 centimeters).

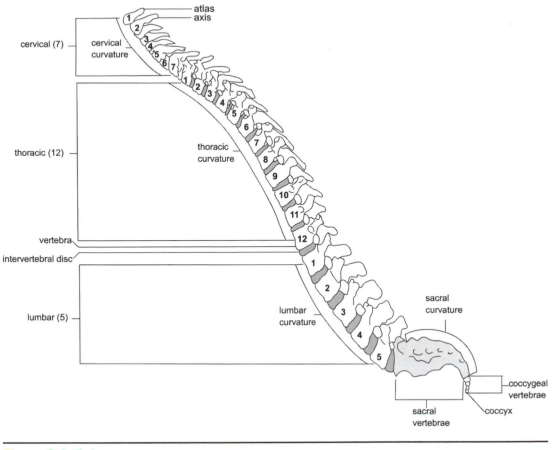

Figure 2.6. Spine.

A TYPICAL VERTEBRA

A typical vertebra has two parts (see Figure 2.7):

- *Body*—the largest part, which is shaped like a short cylinder; this is the forward, or anterior, part
- *Vertebral arch*—a ring of bone formed by paired pedicles, or short, strong processes

Parts of the vertebral arch roof over an opening called the *vertebral foramen*, through which the spinal cord passes. Each vertebra has a spinous process for the attachment of ligaments and muscles of the back. Other processes permit the attachment of muscles and joints that connect individual vertebrae.

The following processes are on the vertebrae:

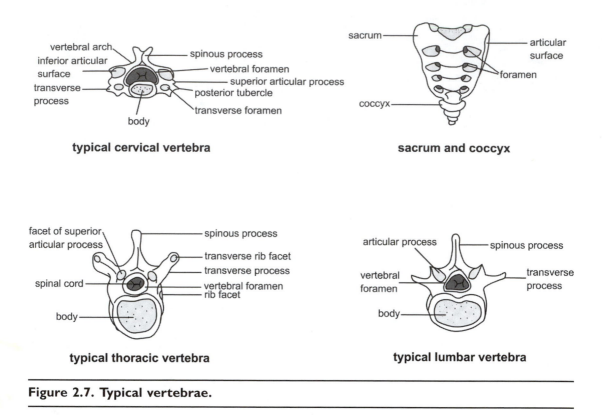

Figure 2.7. Typical vertebrae.

- *Pedicles*—short thick processes that project backwards, one on each side; vertebral notches are located above and below the pedicles
- *Laminae*—two broad plates directed backward and toward the middle
- *Spinous processes*—bony areas directed downward and backward from the junction of the laminae that serve as attachment sites for muscles and ligaments
- *Articular processes*—springing from the junction of the pedicles and laminae, their surfaces are coated with hyaline cartilage, or true cartilage that has a smooth and pearly appearance
- *Transverse processes*—located at each side of the joint where the lamina joins the pedicle; sometimes called *wings*, they also serve as attachments for muscles and ligaments

Because the vertebrae differ in subtle ways in their makeup, they are identified in the following ways:

- The letter C is for *cervical*; T for *thoracic*; L for *lumbar*.
- The vertebrae are numbered from top to bottom. For example, C2 is the second cervical vertebra; T3 is the third thoracic vertebra. Each number begins with the location in the new section.

The Flexible Back

Women in many areas of the world carry tremendous loads on their heads. For example, Bedouin women transport heavy jugs of water on their heads, and Iraqi women balance loads of bricks for construction on their heads.

The flexible back enables them to do these feats. The spine resists such back-breaking loads because the vertebrae can compress. The cartilage disks in between act like tiny balloons filled with water. Like a hydraulic exercise machine that responds to movement by resistance, the spine's powerful hydraulic forces respond to the heavy loads and push them back up, making possible the balancing of impossible-looking loads.

CERVICAL VERTEBRAE

The seven cervical vertebrae in the neck are the smallest of the vertebrae. These vertebrae enable the head to turn in roughly three quarters of a circle without moving the shoulder. Moving the eyes in conjunction with the neck generates almost a full circle of vision. Muscles run from the wings, or transverse processes, on the sides and to the rear of the vertebrae to the skull, shoulder blades, and lower vertebrae. These structures also steady the head on the neck. (See "The Flexible Back.")

Cervical vertebrae are different from thoracic or lumbar vertebrae in that they have a foramen, or hole, in each transverse process. There are also other differences (see Figure 2.6):

- C1 is named for the Greek mythological Titan, *Atlas,* who bore the weight of the world on his neck and shoulder. The atlas, or C1, allows the nodding movement. This bone has no vertebral body, only anterior and posterior arches. It joins with the odontoid process of C2, or axis.

- C2 is the second cervical vertebra, called *axis.* The axis allows movement from side to side. The odontoid process projects above the body and joins the anterior arch of the atlas. The *dens* is a process that forms a pivot around which the atlas rotates to allow head movement.

- C3–6 are vertebrae that are alike but without the special features of C1, 2, and 7.

- C7, called *vertebra prominens,* has the distinctive feature of a long spinal process that is thick and nearly horizontal. (See "The Axial Skeleton in Literature.")

THORACIC, OR CHEST, VERTEBRAE

The thoracic vertebrae become larger from top to bottom to carry additional weight. These twelve vertebrae are intermediate in size between the cervical and lumbar vertebrae. They have the features of a typical vertebra,

The Axial Skeleton in Literature

Aside from the facts of anatomy, the axial skeleton has had a role in great works of literature. After all, the back and the spinal column constitute an expressive instrument that conveys emotions, such as happiness when a person stands erect or sadness when the shoulders droop, as well as attributes of authority or submission. A person's posture and stance can directly express feelings or indirectly symbolize personal shortcomings, social problems, and so on.

In William Shakespeare's (1564–1616) great drama *Richard III*, King Richard III had lived his life with a misshapen back, but his mind and attitude were even more twisted than his body. He detested life and considered that his back deformity was a symbol of evil that his mother had cast upon him in the womb. He hated the enormous mountain on his back that sat in wait to mock his body. His legs likewise were misshapen. He blamed his twisted body for his twisted mind.

The French writer Victor Hugo (1802–1885) wrote a novel, *The Hunchback of Notre Dame*, that featured Quasimodo, a gentle spirit inside a grotesque body. The "hunchback" was possibly a victim of abnormal kyphosis, a skeletal condition, which made him so drawn over that he had a huge hump on his back. He lamented that he looked like a giant whose body had been broken up into pieces and then put back together in an ill-fitting way. Hugo used the misshapen body of Quasimodo to symbolize that appearances are deceiving.

---

plus long slender spines that project downward. The twelve vertebrae join with the twelve pairs of ribs. The ribs join to shallow cups, or costal pits, on the body of the vertebrae. Ten thoracic vertebrae have two costal pits, or **facets**, on each side—one above and one below to join the head of the same rib. This gives extra stability. The two sets move with every breath.

LUMBAR, OR LOWER BACK, VERTEBRAE

These five vertebrae are the strongest and the largest because they support the weight of the body. The transverse processes and neural spine are thicker because they anchor the muscles that twist and level the lower back. Between the vertebrae is a cushion-like disk of cartilage. Although the lumbar vertebrae have features of a typical vertebra, they also have short, blunt spines that project to the rear. The lumbar spines do not overlap, so the area is a good place for a *spinal tap*—a drawing of spinal fluid for the diagnosis of certain conditions, such as meningitis, an inflammation of the covering of the brain called the meninges.

SACRUM

The sacrum is a triangular bone formed by five fused vertebrae. The sacrum makes up part of the pelvis. The fused vertebrae sit like a wedge between other parts of the pelvis.

COCCYX

The end of the vertebral column is the coccyx, or tailbone. It results from the fusion of four coccygeal vertebrae, which are not as complex as the others in that they have no pedicles, laminae, or spine. The coccyx is made from four centers, one for each segment. The forming of the bone occurs in the following order during a person's lifetime:

- Between ages 1 and 4, the first segments ossify, or form bone.
- Between ages 5 and 10, the second unit forms bone.
- Between ages 10 and 15, the third unit forms bone.
- Between ages 14 and 20, the fourth unit ossifies.
- As people advance in age, the segments unite with one another.
- Later in life, especially in women, the coccyx often fuses with the sacrum.

Gymnasts demonstrate the amazing agility of the spine, which is most flexible during the younger years. As people age, knobs of bone grow on the vertebrae and cartilage disks between them become hardened.

Thorax, or Rib Cage

Because the lungs inhale and exhale, they need protection from being damaged. A solid case like the skull would not enable a person to breathe in and out; a better means of protection would be a group of moveable bars forming a flexible cage—just like the ribs. In fact, the ribs are closely spaced with ligaments and muscles in between. They move at the joints, with the spine and breastbone making it possible for them to lift upward during inhaling and to move back down during exhaling.

The thorax, or rib cage, is made of bones and cartilage that protect the important organs of respiration and circulation. Its shape is like an inverted cone, with the narrow part above and the broad part below. From the back, the twelve thoracic vertebrae and the posterior part of the ribs form the cage. The sternum and costal cartilages form the front surface. The sternum is slightly convex and tilts downward and forward. The ribs form its side surface. Eleven intercostal spaces separate the ribs. The term *costal* means "pertaining to the ribs." The intercostal muscles and membranes are also found in this area (see Figure 2.8).

A woman's thorax differs from a man's in the following ways:

- The woman's capacity is less.
- Her sternum is shorter.
- The upper level of her sternum is on a level with the lower part of her body.

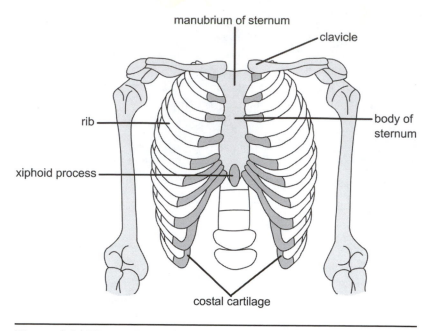

manubrium of sternum

clavicle

rib

body of sternum

xiphoid process

costal cartilage

Figure 2.8. Ribs and sternum.

- Her upper ribs are movable, allowing for a greater enlargement of the upper part of the thorax to accommodate expansion of the uterus during pregnancy.

RIBS

Twelve pairs of ribs are classified as typical, or true, ribs; false ribs; and floating ribs. All three types of ribs have the following structures:

- *Head*—the back and middle ends that join with demifacets, or cup-like structures, of two adjacent vertebral bodies

- *Neck*—a constricted region about $\frac{3}{4}$ inch (2 centimeters) long that is beside the head

- *Tubercle*—next to the neck of the rib, this part joins to the transverse process of a vertebra

- *Body*—the shaft of the rib, the longest part of a typical rib

The ribs are described as follows:

- Ribs 1–7 are true ribs that attach directly to the sternum by means of costal cartilage and a true synovial joint (this joint has a clear lubricating fluid secreted by the synovial membrane; see Chapter 5).

- Ribs 8–10 are false ribs joined via costal cartilage of rib 7.

- Ribs 11–12 are floating ribs that do not articulate with the sternum or costal cartilage of the rib above (see photo).

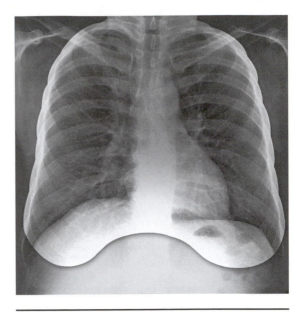

X-ray of normal lungs, showing the rib cage and sternum. © BSIP/Phototake.

Most men and women have twelve complete pairs of ribs. Sometimes a man or a woman will have eleven or thirteen pairs, but this is unusual.

Sternum, or Breastbone

The sternum is a broad, flat bone forming the anterior wall of the thorax. Three parts make up its structure:

- *Manubrium*—the top part of the sternum means "handle," such as that of the handle of a sword.
- *Body*—the middle part articulates with the costal cartilages and ribs 2–7.
- *Xiphoid* process—the lower, or inferior, part of the bone varies in size and joins the bottom of the sternum; its name means "sword shaped."

Ligaments form the alignments to the clavicles, or collarbones, at each side of the top of the sternum. Costal cartilages are bars of true hyaline cartilage made of smooth, tough material that contribute to the elasticity of the rib cage.

The axial skeleton—composed of skull, vertebral column, ribs, breastbone, and pelvic and pectoral girdles—is the main part of the skeleton making up the head and trunk. These eighty bones protect the vital organs of heart, lungs, spinal cord, and brain. They form the foundation for the attachments of the appendicular skeleton (see Chapter 3).

Bones of the Limbs:
The Appendicular System

Whereas the bones of the limbs had a mystical past, they serve today as the creative tools of modern existence (see "Appendages in History"). Bones of the limbs and structures relating to them make up the appendicular skeleton. The word *appendicular* comes from the Latin, meaning "to hang from." Actually, the limbs do hang from the body as appendages.

The axial skeleton of skull, vertebral column, ribs, and sternum becomes the form to which the appendages are hung. Most bones do not lie in the body's central axis but in the extremities. The appendicular skeleton has 126 bones and includes the bones of the arms and legs and those of the shoulder and pelvic girdle.

Mammals, including humans, generally have legs longer than the torso. These bones account for great variation in height. For example, the length of the human spine varies little from woman to woman or from man to man. The woman's spine is basically 24 inches in length, and the man's is 28 inches. Differences in height result from the length of the leg bones. When a group of men are seated, they look the same height; but when they stand up, great differences in height are visible. Arm and leg, hand and foot— these appendages are similar, with the arrangement and number of bones being somewhat alike.

THE PECTORAL GIRDLE

The pectoral, or shoulder, girdle is composed of four bones: two scapulae and two clavicles (see Figure 3.1). Usually a girdle is something that en-

Appendages in History

The bones of the appendicular system have had some unusual uses throughout history. In some cultures, they were sculpted and used as tools. Often they were associated with magic and superstition.

Ancient artisans considered the bones of the limbs ideal for common tools and even bowls and utensils. In the secrecy of night, medieval burglars used bones to pry open tight places and cumbersome locks.

Bones became a symbol of death and a sign of evil. Magicians would render people deaf, mute, and blind when waving bones that were believed to put their victims under death's shadow. In medieval Ruthenia, now a part of Russia, thieves would take marrow from a tibia (shin bone), fill the shaft with tallow, and make a grisly candle. Holding aloft the flaming torch, they would circle a house three times, striking terror into the hearts of the occupants and hoping to put them in an eternal slumber.

Other burglars would cut holes in a leg bone and make a flute that would stupefy their victims with haunting melodies. Less sophisticated evildoers would simply throw a bone on the doorstep, hoping the mere presence of the bone would torment the victims into submission.

---

circles as a complete ring. Looking down at the pectoral girdle from above, one can see a double crescent that encircles the upper part of the body. However, this girdle is incomplete, with the clavicles, or collarbones, being separated by the sternum in the front and a gap between two scapulae, or shoulder blades, in the back. In the back, the scapulae are connected only to the trunk by muscles. The bones allow for the attachment of muscles that firmly bind the arm to the trunk. These muscles also permit free movement of the arms.

Scapulae, or Shoulder Blades

Two pairs of bones in the upper torso connect to the bones of the arm (see photo in color section). Located one in front and one in back, the shoulder blades literally float in a sea of muscles. The name *scapula* comes from a Greek word meaning "to dig." The shape is like a shovel or spade. Early humans used the scapulae of some animals as primitive digging tools. Shoulder blades are difficult to fracture and articulate with the clavicles.

The scapula—a flat, triangular bone that has two surfaces—forms the back part of the shoulder girdle. A spiny, prominent, shelf-like ridge extends obliquely across the blade. It supports the *acromonium process*. This is where the shoulder blade connects with the head of the humerus.

The scapula has two surfaces, three borders, and three angles. One angle, called the *subscapular angle*, appears to be bent in on itself along a line at right angles. This arched form strengthens the body of the bone while the summit of the arch supports the spine and the acromonium. Underneath the clavicle, the *coracoid process* of the scapula projects forward and serves as an attachment site for several muscles (see Figure 3.1).

Clavicle, or Collarbone

The clavicles are located in front, ventral to the rib cage and just above the first rib. One can feel them at the base of the neck where the collar is located. The name *clavicle* is a Latin word that means "little key." To some

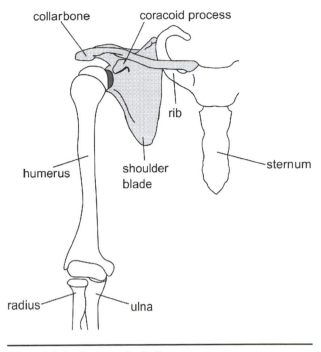

Figure 3.1. Pectoral girdle.

people's imaginations, it looks like an old-fashioned key. The clavicle of birds is very important in folklore. It forms the familiar V-shaped wishbone: Break the long end, and your wish will come true.

The clavicle forms the anterior, or front, portion of the shoulder girdle. Some people describe this long, thin, curved bone as shaped like an S. Others say it is like the italicized lower-case *f*. Placed horizontally at the upper and anterior part of the thorax, it is immediately above the first rib. The upper surface is flat and rough and has impressions for attachments of the deltoid muscle that covers the shoulder prominence in front and the triangular-shaped trapezius muscle covering the back part of the neck and shoulders behind. In fact, the clavicle acts like a fulcrum that enables the muscles to give lateral motion to the arm. (See Chapter 1 for a discussion of fulcrums.)

The part of the clavicle that joins the scapula is triangular. The other end joins a flattened projection of the scapula called the *acromonium process*. This process can be felt as the slight bony projection on the upper surface of the shoulder.

Women have shorter, thinner, less curved, and smoother clavicles than men. In people who perform manual labor, the clavicles are thicker, more curved, and have prominently marked ridges for muscular attachments. The

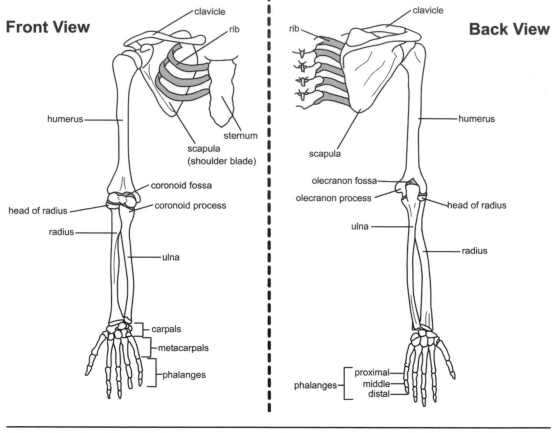

Figure 3.2. Shoulder girdle.

clavicle is the most commonly broken bone in the body because it transmits forces from the arm to the trunk (see Chapter 14).

The Arm

Hooked onto the pectoral girdle are the bones of the arm (see Figure 3.2). Each arm consists of three principal bones: humerus, ulna, and radius. Because of the joints in the bones, the arm is able to move. Also, arms are divided into three segments: upper arm, lower arm, and hand.

HUMERUS

The longest and largest bone of the upper arm is divided into a body and two extremities, or ends. This long bone of the upper arm extends from the shoulder to the elbow. The word *humerus* comes from the Greek word meaning "shoulder." Sometimes the humerus is called the *brachium*. However, it should not be confused with the word *humorous*, which means "comical" or "funny." The common term *funny bone* comes not from a bone but from a nerve—the ulnar nerve—that passes over the elbow and creates a tingling sensation when hit.

The humerus has a smooth, rounded head that articulates with the scapula. On the lateral, or side, surface is a roughened area. Muscles and ligaments attach here. There are also several sites along the shaft of the bone where other muscles are attached. The bones of the forearm articulate on smooth surfaces at the lower end of the humerus.

Two knob-like projections—one on the lateral side and one in the middle—are attachment sites for the common extensor and the common flexor tendons that help move the forearm. Inflammation of the extensors causes the condition known as "tennis elbow." The distal end of the medial **condyle** of the humerus is called the *trochlea*, meaning "pulley." It articulates with the troclear notch of the ulna, which limits side movement and guarantees a hinge action.

ULNA

The ulna is the longer bone along the back of the forearm. The word comes from the Latin term that means "elbow." The end of the ulna is the bony portion of the elbow. At the upper end of the ulna is the *olecranon process*, a projection that fits into the *olecranon fossa* of the humerus when the arm is extended. These two bones form the joint at the elbow. On the lateral side of the ulna is a shallow notch for the head of the radius. The head of the ulna joins with the radius at the lower ends of the individual bones. The flattened surface at the lower end of the radius enables it to rotate around the ulna.

RADIUS

The word *radius* comes from the Latin word that means "ray," something that radiates outward from a center (originally, the word applied to the spokes of a wheel). Whoever named this bone thought this section of the lower arm seemed straight enough to be a spoke in a wheel.

The radius pivots on its long axis and crosses the ulna. Its proximal end has a smooth, rounded surface that articulates with the ulna. Next to the head is the neck; then comes the body, the long slender mid-portion also known as the shaft, or **diaphysis**. The bone ends with the *styloid process*, a projection that joins with the carpal bones of the wrist.

The radius and ulna form the forearm. When viewed from the correct anatomical position, the radius is on the side away from the body (lateral side) and the ulna is on the side toward the body (medial side.) The ulna is longer than the radius and connects more firmly to the humerus. However, the radius is more involved in movement of the hand. A broken arm usually involves this bone.

The Hand

The hand has three parts: wrist, palm, and fingers. Twenty-seven bones form this distal end of the limb. The large number of bones allows for the hand's versatility. A single bone would create a flipper-like, inefficient oar; a single line of bones would lead to bending as a unit, as in the spinal column. But the number of hand bones that spread out in two planes—length and width—introduce two-dimensional flexibility that permits delicate maneuvering. Thus the hand is a superb manipulative organ with four limber fingers and an **opposable thumb** that can act as pincer, grasper, twister, bender, puller, pusher—and manipulator of piano and computer keys.

CARPALS

The wrist, or carpus, has eight small bones. The Latin word *carpus* means "wrist." Ligaments hold these bones tightly together. To allow flexion, or movement, of the wrist, the carpal bones are arranged roughly in two rows. On their inner, or palmar, surface are attached some of the short muscles that move the thumb and little fingers.

Together the bones of the wrist, form a box-like structure. Each of the bones of the wrist has a special appearance and function. Early anatomists named them according to objects that were familiar at the time. The bones in the first, or proximal, row from side to middle are as follows:

- *Scaphoid*—sometimes called *navicular* from the Latin word that means "boat shaped." This bone is located on the floor of the anatomical wrist box. Hyperextension, or excess bending, of the wrist may fracture this bone.

- *Lunate*—from the Latin word meaning "moon shaped" or "crescent shaped"; the second carpal bone in the proximal row.

- *Triquetrum*—from the Latin word meaning "three cornered." The bone is the most medial in the row.

- *Pisiform*—from the Latin word meaning "pea shaped." This **sesamoid bone** is in a tendon and articulates with the triquetrum.

Bones in the second, or distal, row are as follows:

- *Trapezium*—from the Latin word that means "saddle" or "swing." This bone forms a saddle joint with the metacarpal bone of the thumb. Literally, the thumb swings on the trapezium.

- *Capitate*—from the Latin word meaning "head." This largest carpal bone is named for its rounded head. A punching blow with the fist generates forces that are transmitted through the third metacarpal bone to the capitate to the radius.

- *Hamate*—from the Latin word meaning "hooked." This describes the shape of the bone.

METACARPALS

The five metacarpal bones articulate with the bones of the wrist. The word *metacarpal* is from the Latin word that means "after the wrist." The other ends of the metacarpals are rounded and connect with the bones of the fingers. They are embedded in soft tissue and form the palms of the hand. The metacarpals are visible in the back of the hand. They are numbered 1 through 5 beginning with the thumb. Numbers 2 to 5 are almost parallel and do not move, whereas number 1 is set at an angle and has limited mobility.

The metacarpals are described as follows:

- *Metacarpal 1*—shorter and stouter than the others; diverges to a greater degree from the carpus; its surface is directed toward the palm
- *Metacarpal 2*—longest of the five bones; forms a prominent ridge
- *Metacarpal 3*—a little smaller than number 2.
- *Metacarpal 4*—shorter and smaller than number 2
- *Metacarpal 5*—has only one facet on its base.

The metacarpals articulate with the carpals and phalanges of the fingers.

PHALANGES

The fourteen bones of the fingers are called *phalanges* (see photo). One of these bones is called a *phalanx*. In Greek history, the phalanx was a close formation of soldiers marching side by side and front to back. Fingers 2 to 5 have three phalanges decreasing in size from proximal to distal. The thumb has two phalanges. Each finger has a formal name. The thumb is called *pollex*, from the Latin term meaning "strong." It is stronger than the others. (For example, people push a tack into a board with the thumb—hence the term *thumbtack*.) The second finger is the *index finger*, from the Latin word meaning "pointer." The middle finger is called *medius*, from the Latin for "middle."

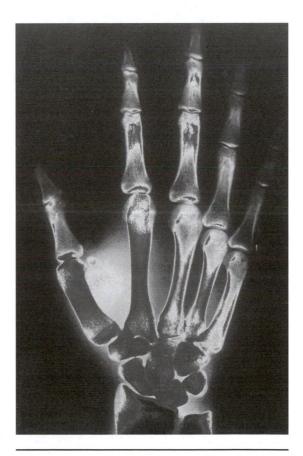

Digitized x-ray of a normal human hand. © Custom Medical Stock Photo.

The fourth finger is the ring finger, or *annularis*, from the Latin for "ring." And last as well as least is the *minimus*, which means "least."

Fingers 2 to 5 have three phalanges, whereas the thumb has only two. The fingers move in and out or up and down. As each phalanx meets, it forms a hinge joint that enables part of the finger to bend. The thumb is more flexible than the other four fingers, with the distal end of its corresponding metacarpal bone being more rounded. Thus the thumb may cross the palm of the hand. This is why it's called an *opposable thumb*. Occasionally, small sesamoid bones are found within the tendons of the hand.

THE PELVIC GIRDLE

Just as the pectoral girdle anchors the arms and hands, the heavier and stronger lower hip girdle supports the structures of the legs and feet (see Figure 3.3). The word *pelvis* comes from the Latin word that means "basin." The pelvic girdle has two large hipbones attached to each other. In the back, these connect with the sacrum to make a bowl-shaped circle of bone. An oddity in the animal kingdom, humans' rounded pelvis is quite different from the elongated pelvis of other animals.

Hipbone

It is a common mistake to think that the hipbone is the large ridge just below the waist. Actually, the hip is at the top of the thighbone and is hidden beneath heavy layers of muscle.

The hipbone is a large, flattened, irregular-shaped bone constricted in the center and expanded above and below. The prominent bones of the hip are called *coxals* or *ossa coxae*, from the Latin words meaning "hipbone." Sometimes the name is *innominata*, from the Latin word meaning "no name." The bones meet in the middle line in front and form the sides and anterior walls of the pelvic cavity.

Each hipbone is composed of three pairs of bones: the *ilium*, from the Latin for "groin"; the *ischium*, from the Greek word for "hip"; and the *pubis*, from the Greek word meaning "adult." The midline divides the bones, with one on each side. Together, they constitute the bony structure known as the hips.

ILIUM

This fan-shaped bone forms the lateral, or side, prominence of the pelvis. Divided into two parts, the body of the ilium forms two-fifths of the *acetabulum*, the socket that fits the ball-shaped top of the thigh. The crest of the ilium can be felt on each side of the body just below the waistline. This is the bone most people mistake for the hipbone.

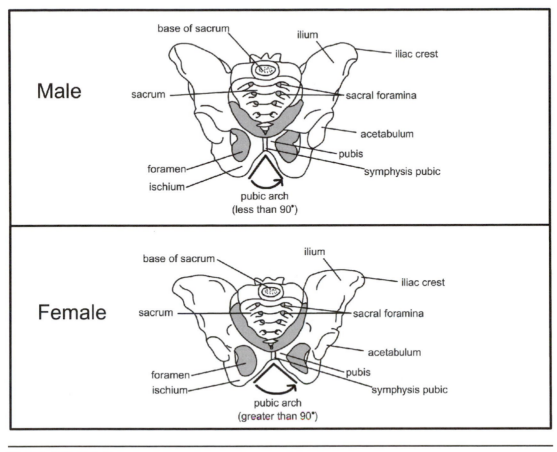

Figure 3.3. Pelvis.

ISCHIUM

This V-shaped bone forms the lower and back part of the hipbone. The body of this bone forms another two-fifths of the acetabulum. It is the site of attachments of many membranes and ligaments, including the hamstring, a major muscle in the groin. The muscles that enable one to sit are also attached to the ischium.

PUBIS

This angulated bone forms the front part of the pelvis and makes up one fifth of the acetabulum. It joins the ischium to form a pair of large holes. The holes, which are prominent in pictures of the skeleton, are called *obturator foramina*, Latin for "stopped-up holes." In reality, a membrane covers this area.

The term *pubis* is derived from the same root as the word *puberty*, describing sexual maturity. One of the signs of such sexual change is the growth of adult hair, or *pubic hair*. The bones located just under this region share this name.

In the front, the pubic bones are joined by cartilage similar to that in the vertebrae. This linkage is called the *pubis symphysis*, Greek for "growing together." In the back, the two bones that form the ilium do not meet but join the sacrum to form the *sacroiliac joint*, the articulation of the hipbones with the sacrum. Five sacral vertebrae form the *sacrum*, a connection so firm that sometimes the area is considered one bone—the sacroiliac. This area sometimes becomes a problem for human beings because of the stress of standing upright on two feet.

The sacrum and the two bones known as *os coxae* form the complete bony structure called the *pelvis*. The rounded, basin-like structure is found only in human beings. However, this arrangement is not perfect, as the pelvic basin tips forward and is not entirely upright.

During birth and early childhood, each coxal bone has the three separate parts of ilium, ischium, and pubis. But by age 20 these bones are firmly fused. The fusion takes place at the large cup-shaped cavity called the *acetabulum*, located near the middle of the outer surface of the bone. The acetabulum forms a socket, or depression, that gives it this name. The structure looked like a round cup the Romans used for vinegar, or *acetum*.

Forensic scientists and anthropologists who examine skeletal remains use the hip for distinguishing between man and woman. In the woman, the bony ring of the pelvic girdle must be large enough for an infant to pass through during childbirth. The average lighter and thinner female girdle is 2 inches wider than that of the male. However, the rest of the woman's skeleton is much smaller and lighter. In the man the bones are more massive, and the iliac crests are much closer.

Where the pubis bones meet at the **symphysis**, a disk of cartilage, in the woman, the bones form a right angle of 90 degrees whereas the man's has an angle of 70 degrees. During childbirth the fetus must pass through the pubic area. If the opening is too small, a problem may occur with birth.

By looking at skeletal bones, scientists can determine the number of children a woman has had because a record of the births is evident in the pelvis. *Parturition* (meaning "childbirth") scars begin to be deposited at about the fourth month of pregnancy when a hormone is released that softens the tendons that hold the pelvic bone together. These scars appear on the back side of the pubis symphysis.

The pelvic girdle, along with the sacrum, is a massive and rigid ring that is very different from the light and mobile shoulder girdle. But the pelvis, while sacrificing mobility, does provide strength and stability. The pelvic girdle supports the weight of the body from the vertebral column. It also protects the lower organs—including the bladder and the reproductive organs—and the developing fetus during pregnancy.

Thigh and Leg

Simply standing on a corner waiting for a bus is in fact an intricate balancing act. Some animals, such as bears, can stand on their back legs for a short period but will soon topple over. Humans, however, can stand upright for hours, leaving their hands free for other things.

The lower extremity is composed of the bones of the thigh, leg, foot, and patella, commonly known as the kneecap. Bones in the legs reflect a compromise with the upper part of the body: They are much stronger than the bones of the arm but are able to move less.

FEMUR

Between the hip and knee is a single bone called the *thigh*, or *femur*. It is the longest and strongest bone in the human skeleton, making up about two-sevenths of the person's height. *Femur* is from the Latin word meaning "thigh."

The parts of this bone include the head, neck, trochanter, body, and patellar surface. The head of the femur, the proximal end, is rounded with ridges that anchor powerful leg muscles. It joins the ilium at an indented surface called the *acetabulum*, fitting into this rounded socket as a ball and socket. Below the head of the femur is the neck, a constricted area that is next to the head. Most of the blood supply to the head streams along this surface. The main shaft of the thighbone forms a wide angle. The neck region tends to become porous over time, especially in older persons, and therefore is a common site of fracture (see Chapter 14).

The lower end of the femur expands into a large flattened area with two bony processes on each side. Jutting out from the junction of the neck of the shaft is the trochanter. The greater trochanter is the insertion point for several muscles, including the *gluteus minimus*. The lesser trochanter projects from the back middle surface and is an insertion point for major tendons and the muscles that form the posterior of an individual. The body is the long slender shaft that ends in the *condyles*, three rounded inferior ends that terminate at the kneecap.

PATELLA, OR KNEECAP

Differing from a similar structure in the arm—the elbow—the knee has a separate bone: the kneecap, or patella. The name of this small, flat, triangular bone comes from the Latin, meaning "small pan." The larger bone of the lower leg, the tibia, joins with both processes on each side. The larger bone of the lower leg joins with the tibia to form the knee joint. The kneecap protects this important joint, which is constantly pushed out ahead of the body as one walks or runs. Like the hyoid bone, the patella is a flat sesamoid bone located just in front of this joint. It is not directly connected to any other bone. The patella develops in the tendon of the front thigh muscle. A ligament attaches the patella to the tibia.

When leg muscles are relaxed, the patella is not held in place by these muscles. The patella moves. Lodged in the tendon, it straightens the leg at the knee and increases the muscle's leverage while protecting the knee.

TIBIA, OR SHINBONE

Like the middle part of the arm, the leg has two bones. Whereas the bones in the arm are nearly equal in length, those in the leg are unequal. On the side nearest the body (medial side) is the larger of the two bones—the shinbone, or tibia. It is the second-longest bone in the body. The word *tibia* comes from the Latin word meaning "flute," which its length and shape resemble. The shinbone is long and skinny; it is slim where the stresses are least.

The tibia is the weight-bearing bone of the leg. It possesses heavy prominences, or condyles, that articulate with the femur to form the knee joint. In the front of the lower leg, the tibia can be felt as not very smoothly rounded and with a protruding ridge. At the distal end at the ankle, the heavy bony protuberance it forms at the inside of the ankle can be felt.

FIBULA, OR CALFBONE

This bone is much thinner than its partner, the tibia, and does not carry the body weight. It mostly anchors muscles that move the foot. The name comes from the Latin word meaning "pin." Its relationship to the tibia resembles the pin on the back of a brooch: well hidden. The fibula cannot be felt, as it is securely embedded in muscles. Sometimes it is called the *splinter bone*, as it resembles the splinter off the shinbone. The end of the fibula can be felt as the bony prominence on the outside of the ankle.

The Foot

The foot, or pes, consists of the ankle, instep, and five toes. Twenty-six bones make up the distal limb (see Figure 3.4).

TARSUS, OR ANKLE

The ankle has a series of seven irregular bones, one less than the eight of the wrist. The word *tarsus* is from the Greek term meaning "wicker basket," apparently suggesting that the separate bones resembled the interwoven wicker strands of a basket. Bones of the tarsus are as follows:

- *Calcaneus*—from the Latin word meaning "heel." This largest tarsal, the strongest bone in the foot, extends back to form the heel. By extending backward, the two heels along with the two soles of the feet give humans **bipedal** support.

- *Talus*—from the Latin term meaning "ankle." This ankle bone articulates with the tibia, fibula, calcaneus, and navicular bones.

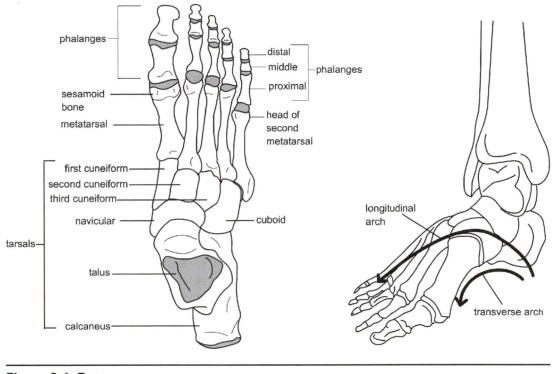

Figure 3.4. Foot.

- *Navicular*—from the Latin word meaning "boat shaped." This bone articulates with the head of the talus and all three cuneiform bones.

- *Cuboid*—from the Latin word meaning "cube shaped." Soldiers in Rome used these bones, usually from horses, as dice for gambling.

- *Cuneiform*—from the Latin word meaning "wedge shaped." There are three of these bones.

As in the wrist, the bone structure enables the foot to move up and down in a single plane. The calcaneus helps support the weight of the body and serves as an attachment site for muscles of the calf of the leg. In fact, when standing, the body weight presses down through the talus and is divided evenly: Half travels to the heel bone, and the other half goes to the five remaining bones and the arch. Wearing high-heeled shoes upsets this balance, throwing the body weight to the balls of the foot.

METATARSALS

The foot contains five metatarsals that join five sets of phalanges. The five metatarsals are similar to the metacarpal bones of the hand. Each of the metatarsals articulates with at least one of the tarsal bones, and sometimes with other metatarsals. The foot is shaped to form two main arches where the metatarsals join with the tarsal bones. One of the arches is longitudinal, lying perpendicular to the transverse arch (see Figure 3.4). Together, they strengthen the foot and act as a spring to cushion certain movements. From any of a variety of conditions—including poor prenatal nutrition, excessive weight, fatigue, or incorrectly fitted shoes—lowered arches, or flat feet, can result. This problem causes unnatural stress and strain on the muscles of the foot and may lead to fatigue and pain while walking.

PHALANGES

The bones of the toes are similar in number to those of the fingers. Four of the toes have three phalanges, and the great, or first, toe has only two. Just like the fingers, the first digit of the big toe has two bones. The name for this bone is *hallus*, from the Latin word meaning "big toe." However, unlike the fingers, the toes have a sturdy, stout shape that bears weight. The bones help the foot push off, aiding in balance as they firmly grip the ground.

The importance of the human appendages is underlined by specialization. Quadripeds—animals that walk on all fours—have similar structures in all limbs. Apes and monkeys have hind feet that are very much like hands. In addition to opposable thumbs, these animals have opposable big toes that enable them to grasp. In humans, the limbs have more specialized jobs. The hands function to create and to work; the feet support.

The appendicular skeleton has 126 bones and includes the bones of the arms and legs, as well as those of the pectoral and pelvic girdles. The structures of the muscles, tendons, ligaments, and cartilage surround the bones at the joints and enable movement.

Joints, Ligaments, Tendons, and Cartilage

Not only is the skeleton a framework for the body, but it also constitutes a movable machine. Bones themselves are rigid, so the only possible motion occurs where two bones come together.

Indeed, the entire body is a complex interaction of matter and motion. A new field, biomechanics, merges the human machine and mechanics (see Chapter 13). Replacing joints diseased from arthritis, regrowing bone, and bone tissue engineering are hot research topics. These are only some of the current and future efforts to fathom the complex engineering of the human machine. (See "Puppets and Joints.")

The sophisticated name for joint is *articulation*, from the Latin word meaning "to join." The study of the joints is known as arthrology, from the Greek word *arthro*, meaning "joint." The same root word is found in **arthritis**, meaning "inflammation of the joint."

Just because a joint is present does not imply mobility. Some joints allow no movement, some permit minor movement, and some allow free movement. The joints may or may not have ligaments that attach bone to bone or bone to cartilage.

Custom-designed according to the body part, joints are divided into three classes that describe the amount of movement between bones:

- **Synarthroses**, or immovable joints, permit no movement.
- **Amphiarthroses** permit only slight movement.
- **Diarthroses** are freely movable joints.

Puppets and Joints

Although the scientific study of joints and their role in movement is a recent development, people throughout history have marveled at how the joints enable a body to move. The principle of how joints work was imitated in making puppets. By looking at how their own joints worked, ancient artisans designed puppets called *marionettes* that would move when attached strings were pulled.

Puppetry is an ancient form of entertainment that is still popular in many countries. Puppets have been found in the toys of ancient Greece and Rome. In fact, the Greek philosopher Aristotle (380–322 BCE) wrote that movement of animals is like that of automated puppets. "For they have functioning parts that are of the same kinds: the sinews and bones. The bones are like pegs ... the sinews like the cables" (*De motis animalium*).

The classic children's story, *Pinocchio*, tells the story of Geppeto, the puppet maker, who wanted to create a miniature boy. He carved the image out of a block of wood. But something was tremendously wrong. The solid boy would not move. Looking at his body, he created individual wooden bones like the ones in his arms and legs and attached strings at the right places to pull the parts. Pinocchio, the little puppet who desired to become a real boy, was born.

Likewise, one solid piece of bone would not move. Nature has solved the problem in all vertebrates by dividing the skeleton into many bones and creating joints where the bones come together. However, as Geppeto found out, no man-made structure can approach the extraordinary device that nature has created in joints.

Tough **collagen** fibers called *ligaments* (from the Latin word that means "to tie") bind joints together and link bone to bone. Many are named from the two bones they attach. For example, the *sphenomandibular ligament* attaches the spine of the sphenoid bone and the mandible; the *stylohyoid ligament* connects the styloid process with the horn of the hyoid bone.

Ligaments and tendons are like rubber bands that hold the musculoskeletal system together. The material is very strong in resisting heavy loads. Although tendons and ligaments are made of dense, fibrous connective tissue, they differ in makeup and function. Ligaments bind one bone to another; tendons bind muscle to bone. Ligaments are 55 to 65 percent water. A special type of protein called *collagen* makes up 70 to 80 percent, with the protein *elastin* making up 10 to 15 percent of the dry weight (minus water). Tendons have 75 to 85 percent collagen and less than 5 percent elastin. Generally, tendons have fewer cells than ligaments do. And they are very sturdy. For example, the tendon of the foreleg of a horse can support the weight of two large automobiles. Sometimes a tendon is called by the Anglo-Saxon word *sinew*, which means "tough."

IMMOVABLE, OR FIBROUS, JOINTS

Immovable joints are called *synarthroses*, from two Greek words: *syn*, meaning "together," and *arthro*, meaning "joint." These joints are firm in their positions to prevent gliding or sliding. Examples are sutures in the skull.

Sutures are limited to the skull. Instead of cartilage between bones, fibrous tissue is located there. Necessary for skull growth, the joints are well marked in the young skull and barely visible in the aging skull. The only movement in the area is at birth, when cranial bones overlap to allow the baby to pass through the birth canal. Serrated little teeth that fit together join the suture. Later in life when growth is complete, they fuse. Suture joints in the skull are the following:

- *Coronal sutures*—the articulation between the frontal bone and the two parietal bones

- *Intermaxillary suture*—the midline of the hard palate that marks the line of the two palatine shelves

- *Lambdoidal suture*—the joint between the occipital and parietal bones that resemble the Greek letter lambda (Λ)

- *Metopic suture*—a midline suture forming the articulation between two centers of the frontal bone

- *Pterion suture*—joins four bones: the greater wing of the sphenoid, the frontal, the parietal, and the squamous part of the temporal bone; this area is easily fractured with a blow to the side of the head

- *Sagittal suture*—joins the parietal and squamous portions of the temporal bone

CARTILAGE, OR AMPHIARTHROSES, JOINTS

These fibrous and cartilagenous joints occur where two bones are separated by a material that gives a little. Three such joints are synchondrosis, gomphosis, and syndesmosis.

Synchondrosis

This common type of joint is named from two Greek words meaning "join together with cartilage." The structure is a cartilage "sandwich" with bone as the "bread" on each side. The bone and cartilage fit together perfectly, and the whole joint is shaped like a cup. If movement occurs the growing bone will be damaged, causing a condition known as a *slipped epiphysis*. (The **epiphysis** is the portion of a bone that is attached to another bone by a layer of cartilage.) A long pin can be inserted to hold the joint in place.

Gomphosis

These peg-and-socket joints occur between the teeth and jaws. The periodontal ligament holds this joint, which gives only a little. When the teeth bite down on a hard piece of candy, this joint absorbs the shock.

Syndesmosis

This type of joint is commonly known as a tight joint. In fact, tight ligaments limit many joints. An example is the inferior tibio-fibular joint between the two lower leg joints. A tight ligament limits the movement of this joint.

Symphysis is a word that describes two bones united by cartilage but designed to give a bit. For example, the symphysis pubis with ligaments and cartilage is normally closed, but female hormones signal it to open for childbirth.

The disks between the vertebrae are made up of fibrous cartilage and form a symphysis. These disks, which give just a little, are important shock absorbers between vertebrae. Joints between the disks of cartilage and other tissues permit some movement.

What Is Cartilage?

Cartilage comes in different types, each suited to a particular use:

1. *Hyaline cartilage* is the most prevalent type. It is found at the ventral ends of ribs, in the rings of the windpipe, and covering the joint surfaces of bones. The covering at the joint surfaces of bones is called *articular cartilage*.

2. *Elastic cartilage* is found in the external ear, the Eustachian tubes in the middle ear, and the epiglottis—the flap that covers the windpipe during swallowing. Compared to hyaline cartilage, it is more opaque, flexible, and elastic. It is yellow and very dense.

3. *Fibrocartilage* occurs in the disks between vertebrae, in the pubis symphysis in the pelvis, and in the bony attachments of certain tendons. It may also form when hyaline cartilage is damaged.

Cartilage is mainly collagen embedded in a firm gel. Collagen is an albumin-like protein in connective tissue, cartilage, and bone. The material is more flexible than bone and lacks blood vessels. Cartilage cells receive nutrients from the diffusion of fluids from nearby capillaries (minute blood vessels that connect the smallest arteries and veins). Nose and ears are examples of cartilage that deteriorate at death. This is why a skeleton never has a nose or ears.

SYNOVIAL, OR DIARTHROSES, JOINTS

Synovial joints are very different from fibrous joints in that they are constructed to allow a range of motion. Yet they are sturdy enough to hold the skeleton together. Cartilage covers bones near synovial joints so that ligaments may attach.

Constructed to give power and motion, the ends at the synovial joints have a thin but tough layer of articular cartilage. This clear material lessens friction and cushions joints from jolts. If the coating is destroyed, the bones grind against one another, producing a creaking sound. Between the bones and at the center of a synovial joint is the joint cavity that gives bones some freedom of movement. Synovial joints occur in a range of sizes and shapes (see Figure 4.1).

Ball-and-Socket Joint

These joints are in the shoulder and hip. They allow for the freest movement. The surface of the epiphysis, or rounded head, of one bone fits in the cup-shaped socket of the other. For example, the ball of the femur fits tightly

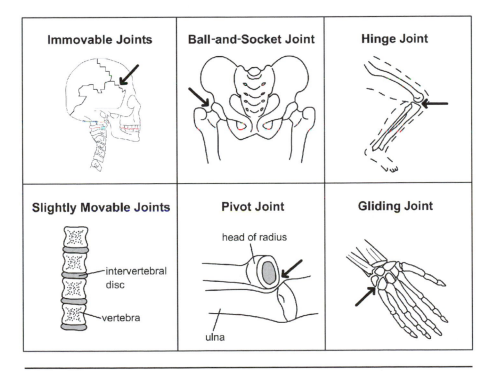

Figure 4.1. Joints.

into the acetabulum of the hipbone. A rim of cartilage lines the socket and aids the firm grip on the femur. Some of the strongest ligaments in the body reinforce this joint. Because of the ball-and-socket joint, a person may move the leg in almost any position. **Dislocation** injuries are most common when the knee is flexed, as when sitting in a car and the impact from a collision causes the knee to strike the dashboard.

A similar ball-and-socket joint located between the humerus and scapula permits an even greater degree of freedom. The socket here is shallower than the one at the hip joint, enabling a person to turn the arm in a complete circle at the shoulder. For example, baseball pitchers doing a complex windup can move their arms in a full circle.

Hinge Joints

As a door swings back and forth on hinges in one plane only, so do bones connected at hinge joints. There are forty hinge joints in the human body, including the elbow and knee.

The elbow allows motion in one plane only. This complex synovial joint consists of humerus and radius, humerus and ulna, and radius and ulna articulations—all within a common articular capsule. The proximal epiphysis of the ulna just fits between the two epiphyses of the distal end of the humerus, allowing movement back and forth but not from side to side.

More complex is the body's largest joint, the knee. This joint links the two rounded bulbs of the femur with the condyles of the tibia. The patella, or kneecap, covers this joint. In the act of walking the knee is constantly rolling, gliding, and shifting orientation. To bear these enormous loads, a host of ligaments and tendons go into action. Two ligaments on each side of the joint prevent it from moving too far to one side. When torn, these ligaments usually repair themselves.

Not so with a second group of ligaments strapped across the joint, the cruciate ligaments. These **intracapsular ligaments** are the anterior and posterior cruciate ligaments. The first—sometimes referred to as the *anterior cruciate ligament*, or ACL—is infamous in sports medicine. The role of these ligaments is to stop the joint from moving too far backward or forward. When the ACL is damaged, the knee is in great peril because it is extremely difficult to repair. Damage to the ACL has put many athletes out of their game.

The joint capsule of the knee has ligaments as well as menisci to add to stability. The medial and lateral menisci are crescent-shaped wedges of cartilage with a smooth, slippery surface. Menisci enable the joint to glide easily and to absorb the shock of daily activity. Menisci are the knees' weakest link, accounting for 90 percent of all knee surgeries. Muscles, such as the *vastus medialis* that hold the knee in place, also add to the stability of the region.

Other hinge joints occur between the first and second phalanges of the fingers and thumbs and between the second and third phalanges of the fingers. The same is true for joints in the toes. The lower jaw, or mandible, is mostly a hinge joint, but it can move from side to side in a rotary motion.

Pivot Joints

These joints provide rotary movement in which a bone rotates on a ring or a ring of bone rotates around a central area. In shaking the head "no," the movement between the first two vertebrae allows the turning of the skull on the spine.

Gliding Joints

These joints are found between the carpals of the hand and the tarsals of the feet. The joining parts slide over each other with angular or rotary motion. Similar joints are in the ribs and vertebrae.

Angular Joints

Where a football-shaped bone fits into a concave cavity, an angular joint occurs. This is found in the wrist, and it permits movement in two directions. Sometimes these are called *condyloid* or *ellipsoid* joints.

Saddle Joints

Similar to an angular joint in its range of movement, this joint is found only in the thumb. Each bone that forms a saddle joint has a concave and convex articular surface.

Plane Joints

These joints occur between two flat bones where one moves horizontally over the other in both directions. They are found in the hand.

SYNOVIAL FLUID

When the bones move against each other, friction must be reduced. Several features work together to accomplish this. Portions of the bone are lined with a smooth layer of cartilage, and a capsule called the *synovial capsule* holds the bones together. This joint capsule permits movement and has great strength to prevent dislocation. Lining the joint is a membrane that secretes a lubricant called **synovial fluid**. Like egg white in texture, the fluid contains a substance called *hyaluronic acid* that helps to ease friction.

In these joints there is also a membrane sac called a **bursa**. Bursae produce synovial fluid that bathes the end of the bone, allowing fluid movement. If the bursae become inflamed, a condition called *bursitis* causes severe pain (see Chapter 15).

Because bones are rigid, the only motion occurs where two bones come together as joints. Ligaments bind one bone to another; tendons bind muscle to bone. Although all joints do not move, these structures at the joint allow the human being to move.

Bone Growth and Renewal

Beginning with the famous Roman surgeon Galen (129–216), traditional wisdom held that bone was just like stone—inert, formed, and nonliving. However, in the eighteenth century a French scientist, Henri-Louis Duhamel (1700–1762), heard of an experiment showing that red dye fed to animals would make their bones red. If bones were stone, this could not happen, Duhamel reasoned. He then fed his own chickens, turkeys, and pigs ground parts soaked in madder, a red dye, and found that their bones became red on the outermost layers—the areas presumably formed most recently. Under the red, the bone was white. By alternating a regular diet with the red madder diet, Duhamel found the bones had alternating layers of red and white. Later, scientists discovered that bones grow by adding layer upon layer, a process called **accretion**.

STRENGTH OF BONE

Bone is one of the hardest structures of the animal body. Possessing both toughness and elasticity, it is pinkish-white externally and deep red within. Engineers estimate that 1 cubic inch of bone can withstand a load of about 19,000 pounds, the weight of five small pickup trucks. Fractures of bone, such as the femur, happen when it is twisted or receives a blow from the side with a pressure of a few hundred pounds per square inch. Bones' strength could be considered four times greater than the strength of concrete.

Bone is light. About 14 percent of total body weight, or approximately 20 pounds, is bone. A steel bar of the same size would weigh four to five times more. Ounce for ounce, bone is stronger than other popular strong materials, including aluminum and steel.

STRUCTURE OF BONE

To understand the structure of bone, one must examine it at three levels:

- Overall view of bone, or its morphology
- Microscopic structure
- Chemical structure

Bone Morphology

Morphology is the branch of study that deals with the structure and form of an animal or plant. Most bones have a major structure, the body, and one or more places where they join with other bones. Long bones join at articular heads and ends. Cranial bones join at complex sutures; ribs join the vertebrae at simple facets. If the bone is healthy, it will appear smooth and dense.

Inside the bone, the internal structure may be either a thick cortex that is variable in width or a spongy honeycomb (this is called *cancellous bone*). In long bones, the cortex and a very thin layer of cancellous bone form a tube surrounding a hollow *medullary cavity*, which serves as a storage tank for fat and yellow marrow. Flat bones have a similar structure but no medullary cavity (see photo in color section). In old bones the aging process caused the inner bone to disappear, allowing the two opposing cortical surfaces to cave in on each other. Because the strengthening structures are not working properly, fractures may occur.

Compact tissue is always on the exterior of the bone, and cancellous tissue is in the interior and at the ends (see Figure 5.1). Different bones have varying amounts of the two tissues, according to the required strength and lightness of the bone.

COMPACT BONE

Dense like ivory, compact bone has slender fibers and **lamellae** that form a reticular, or maze-like, structure. Cavities are small in compact bone, and solid matter between the cavities is abundant. Blood vessels permeate bone, and a fibrous membrane—the **periosteum**—encloses the bone except at the ends, which are covered by articular cartilage. The cylindrical interior of the long bones is filled with marrow.

CANCELLOUS BONE

From the Latin word meaning "lattice," this type of bone occurs at the ends of long bones and lines the hollow centers. Built of bundles of crystal

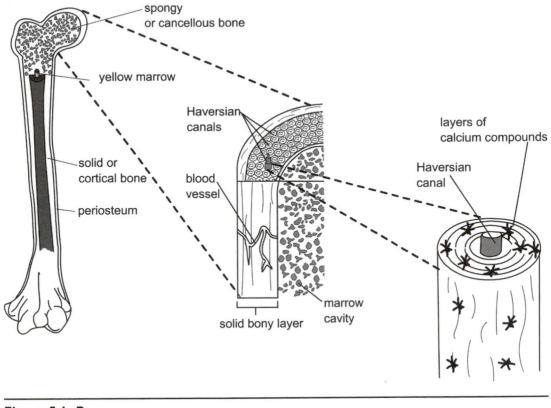

Figure 5.1. Bones.

collagen with no Haversian systems, or microscopic concentric circles, this tissue has a honeycomb of spaces and a smaller quantity of solid matter. Instead, there is a network of struts and braces called *trebeculae*; these resemble coral.

Cancellous bone, much lighter than compact bone, is designed for resistance. The beams of the trabeculae resemble a dome in which the supporting structures occur along lines of greatest stress.

A Microscopic View of Bone

If a section of dense bone is viewed under a microscope, the slide reveals a mysterious landscape (see Figure 5.1). Varying tree-like rings form concentric circles called *Haversian systems*. Some of these structures are as large as 0.12 millimeters in diameter, but the average is 0.05 millimeters. Haversian systems were named to honor the English physician Clopton Havers (1655–1702), who first described them in 1691. Each ring, called a *lamella* is made of fine threads of collagen peppered with mineral crystals of calcium and phosphorus. These may be pictured as sheets of paper pasted one over the other around the central hollow cylinder. The fine transparent fibers can be recovered by treating ground-up bone with acid. Blood, lymph vessels, and nerve filaments form the canal in the very center of the concentric rings (see photo in color section).

On the concentric rings, looking like bubbles in a piece of glass, are the **lacunae** (from the Latin word meaning "small lake"). A single cubic inch of compact bone has over four million lacunae. Each lacuna is occupied by a branched bone cell. Called **osteocytes**, these cells maintain bone tissue. Bone cells do not completely fill the lacunae. In young bones, the osteocytes' branches reach into the surrounding area.

Radiating from the central canal are minute, hair-like channels called *canaliculi*, Latin for "little canals." These tiny structures cross the lamellae and connect the lacunae with neighboring lacunae and also with the Haversian systems.

MARROW

Marrow fills the cavities of the long bones, the spaces in cancellous tissue, and the Haversian systems. A marrow-filled hollow bone is lighter than a solid one. When lightness is important—as in some birds—true hollowness is found. For example, flying birds must conserve weight, so their bones are hollow and fragile. In fact, many birds have feathers that weigh more than their bones. However, in humans and other mammals, few truly hollow bones exist.

Marrow may vary according to the type of bone. For example, in long bones the marrow is yellow, containing 96 percent fat cells. In flat and short bones in the articular ends and in vertebrae, marrow is red, containing 75 percent water and 25 percent other materials.

The following cells are located in the marrow:

- *Myelocytes*—large cells from which white blood cells are made

- *Erythroblasts*—from which the red blood cells of adults are derived

- **Osteoclasts**—giant cells that have many nuclei; found in both kinds of bone marrow, they do their work in bones' small shallow pits, or cavities.

BONE VESSELS AND NERVES

Blood vessels are numerous. The blood in compact bones passes through the tiny openings in the compact tissue running through the Haversian canals that transverse the outer layer of bone. Vessels supply blood to the spongy cancellous tissue.

One large artery supplies blood to the marrow. This artery enters the bone at the nutrient foramen (opening) located at the center of the bone and then perforates compact bone at an oblique angle. This medullary, or nutrient, artery is usually found with one or two veins and sends branches upward and downward. In flat and short spongy bones, one or more of the large holes transmit blood to the central parts of the bone. In flat cranial bones, veins are large and numerous, running in winding canals in the sides of the canals formed by thin lamellae of bone. In cancellous tissue, veins are enclosed and supported by bony material with very thin coats.

Lymph vessels also occur in bone and run in the Haversian systems. Lymph vessels are part of the lymphatic system, a one-way transporting network of vessels that carries fluid to the tissues. Likewise, nerves in the periosteum accompany nutrient arteries into the interior of the bone. Nerves are most numerous in articular extremities of the long bones, in the vertebrae, and in the larger flat bones.

PERIOSTEUM

Wrapping each bone is the periosteum. The word *periosteum* comes from two Greek roots: *peri*, meaning "around," and *osteo*, meaning "bone." It consists of two united layers:

- Outer layer mostly of **connective tissue** with a few fat cells
- Inner layer of fine elastic fibers that form dense networks and are separated into layers

In young people, the periosteum is thick and vascular. It is connected at each end with the epiphysial cavity and loosely on the body of the bones. A layer of soft tissue with a number of **osteoblasts** enables bone to form as a young person grows. Later in life, the periosteum is thinner and less vascular.

If a surgeon peels the periosteum from bone, small bleeding points mark the spots where blood vessels penetrate the bone. Also, fine nerves and lymph vessels accompany the arteries. Because of these blood vessels and nerves, the periosteum is sensitive to injury. Periosteum damage from fractures and bone bruises causes great pain.

CELLS

Looking at the external appearance of bones and the position of each bone in the body gives the impression that bone is static. After all, a hard material makes up 45 percent of its weight, and this part is nonliving. However, as microscopic study reveals, living cells in bones and cartilage make the skeletal system alive and well.

As the skin and other organs are in a state of flux, so is this system. The cell is the unit of this living tissue. In 1665, the Englishman Robert Hooke (1635–1703) looked at a small piece of cork and noted tiny compartments that reminded him of rooms in a monastery that he called *cells*. In Greek, the word *cytos* means "cell." Later scientists found that the animal cell is made up of a nucleus, cytoplasm, and a cell membrane. A *tissue* is a group of cells with the same specialization. For example, bone cells make bone tissue.

When repeating the experiment of Henri-Louis Duhamel, who fed colored madder to animals and then observed their red bones, the British physician John Hunter (1728–1793) was the first to discover how bone renews itself. He too fed colored madder; but he found that when he stopped feeding it, the red bone disappeared. In the 1760s, when most people believed bones were no more alive than stone, Hunter determined that the skeleton is made of complex living tissue that constantly renews itself.

Bone removes worn cells and creates new ones with the help of three kinds of cells:

- *Osteoclasts*, which consume old and worn bone matter. The word comes from the Greek term meaning "bone breaker."

- *Osteoblasts*, which manufacture new bone tissue. The word comes from the Greek term meaning "bone bud."

- *Osteocytes*, which act as caregivers, or nurse cells.

These three cells are the workhorses of bodily metabolism (the sum of all physical and chemical changes in an organism).

Osteoclasts. These giant cells have several nuclei and usually live a very short life. As the skeleton develops, they help bones reach their proper length. After growth ends, they are instrumental in shaping bones and in creating a medullary cavity in the middle of long bones.

Resorption is the term used to describe the breaking down of bone that occurs at different rates throughout the body. In mature bones, osteoclasts may be active on less than 1 percent of the surface at a given time.

The **parathyroid glands,** located beside and behind the thyroid, control the operation of the osteoclasts. A hormone from this gland controls the development from cells carried in the bloodstream. The cells act as teams that burrow tunnels in old bone. They send out tiny projections, called *villi*, that invade and secrete **enzymes** and acids to decimate the old bone. Then the osteoclasts digest the crumbled matrix, sending the minerals into the bloodstream. After they have done their work, they disappear.

Osteoblasts. Next, the osteoblasts go to work. Found just below the periosteum on the outer surface of the bone, they are busy rebuilding on about 4 percent of skeletal surfaces at any time. Damaged bone quickly demands more osteoblasts to lay down new bone. Several months may pass before their work is done. New material soon hardens from the calcium the osteoblasts bring in from the bloodstream.

Osteocytes. Osteocytes, the most stable of the three kinds of cells, are embedded in the hard matrix of the bone and act as caregivers to the other two types. They begin their lives as osteoblasts but become caught in their own secretions, which harden around them. Losing their bone-building power, they occupy the small lacunae, or gaps, connected by the canaliculi. Osteocytes then help to create a flow of liquid calcium and phosphorus through the channels between the matrix and the bloodstream. They orchestrate the blood stability of the bone and the processes of resorption and remodeling. Although they appear to be inactive, osteocytes are a vital part of the system: Their death means the death of the bone.

Chemical Composition of Bone

Bone is an interesting combination of chemicals (Table 5.1). Many school science classes conduct a classic experiment with bone. They put a chicken or turkey bone in a jar with vinegar, a weak acid, for a day or so, after which the bone looks the same as before but now is rubbery and flexible and can be tied in a knot. The acid has taken calcium and phosphorus out of the bone. But in a cross-section, the original arrangement of the Haversian systems is visible.

If the bone then gets put in a high-temperature oven, it will retain its original form but will turn white and brittle. Now about one-third of its original weight, it will shatter with the slightest force. This fragility indicates that the now-missing mineral salts are essential in the work of healthy bones.

Approximately 1.5 percent of a person's body weight is calcium. In the blood plasma, calcium plays a vital role in blood coagulation, cardiac, skeletal, and nerve functions. Phosphorus is required for other vital functions. The mineral in living bone, **hydroxyapetite**, forms a crystal structure. (See "Bones and Coral.")

Small amounts of sodium, magnesium, fluoride, and carbonate are also found. These bone salts are relatively insoluble and usually remain in equilibrium so that the calcium taken in matches the output. Indeed, calcium is constantly being replaced. In infants, there is a 100 percent turnover each year; in adults, there is about an 18 percent turnover due to resorption and reformation of bone tissue. Vitamin D promotes the absorption of calcium.

TABLE 5.1. Chemical Composition of Bone

Compound	Chemical formula	Percentage in bone
Calcium phosphate	$Ca_3(PO_4)_2$	58
Calcium carbonate	$CaCO_3$	7
Calcium fluoride	CaF_2	1–2
Magnesium phosphate	$Mg_3(PO_4)_2$	1–2
Sodium chloride	$NaCl$	Less than 1

Bones and Coral

Living bones and coral, the limestone skeletons of marine organisms that form the famous reefs and islands of warm seas, have much in common. They are both made of the mineral hydroxyapetite, which forms a crystal structure.

Many ancient practitioners used coral, or coral calcium, as a medicine. Five thousand years ago, medicine men of India wrote about its use in their medical books. Arabs realized the importance of coral for health and introduced it as medicine to Western Europe when they conquered the region in the ninth century. In 1685 doctors in Panaranda de Duero, in northern Spain, established the world's first pharmacy, which is now a museum. A vial of coral is displayed along with an inscription that touts its beneficial effects on the heart and mood but warns that only the blond coral should be used for medicine. In Okinawa, an island off the coast of Japan famous for its number of centenarians, natives have taken calcium coral for over 500 years and claim its use contributes to longevity.

Because coral is naturally compatible with the body, orthopedic surgeons have experimented with the use of coral in bone grafts. The body readily accepts coral; muscles and tendons grow into the coral structure as if it were human bone.

If this vitamin is deficient, bone fails to mineralize; as a result, children may develop rickets and adults may develop a condition known as *osteomalacia* (see Chapter 17).

FORMING BONE: OSSIFICATION

The metabolism of calcium and phosphorus salts is involved in the process of bone formation, or *ossification*. The general form of the human skeleton is determined long before birth in the early embryo around the sixth week of gestation. As early as the seventh week, bone cells begin to replace these materials with bone tissue. Bone forms in two ways:

- from fibrous membranes
- from flexible hyaline cartilage

Ossification continues throughout life but is primarily focused in the first twenty-five years. Afterwards, bone tissue is capable of local growth, as seen in the healing of a fracture.

Bones from Fibrous Membranes

In ossification from fibrous membranes, no cartilage precedes the bony tissue. Bones of this type include the flat cranial bones that protect the brain, the lower jaw, and parts of the clavicle.

The membrane that becomes bone is like connective tissue; it is made of fibers and granular cells in a matrix. The outside is fibrous, but the interior contains the bone builders; the osteoblasts. Many blood vessels run in and out of the matrix.

The procedure by which the thick membrane gradually mineralizes is as follows:

1. The bone-building cells, the osteoblasts, create a tiny network of *spicules*, or little spikes, that radiate from the center of ossification. These spicules become fibers that resemble white fibrous tissue.

2. Deposited between the spicules are calcium and phosphorus, mineral salts that look like tiny grains of sand. Some osteoblasts are enclosed in the calcified material.

3. As the grains grow together, the tissues appear transparent and the fibers are no longer visible, but the osteoblasts are still there.

4. The fibers continue to grow, calcify, and create a new network of spicules.

5. Calcification continues, forming an open framework characteristic of spongy bone. Osteoblasts on the surface form the additional fresh layers of bone, thickening the trabecular structure.

6. Layers are deposited here and around the larger vascular channels that become the Haversian systems. Periosteum will soon cover the surface of the new bone.

The skull is an example of bone formed as membrane that makes separate plates. A thin layer of compact bone is produced just under the periosteum, forming a hard covering for the spongy bone underneath. As the bony plates grow toward each other, they meet and form the immovable joints called *sutures*. At birth, the bones of the cranium have not completed ossification; the membrane-covered space is called the *fontanel*, or soft spot. This area where the bones have not met gradually grows together by about age 2.

Bones from Cartilage

Except for the clavicle, the bones of the limbs and the vertebral column are formed from cartilage. Before ossification, bones of this type are a mass of cartilage. For example, in a long bone the process of ossification begins in the center and proceeds to the extremities, which remain cartilage for a period of time.

The transformation to bone begins with the erosion of cavities in the cartilage. Blood vessels invade the cartilage, and osteoblasts—the bone builders—begin their work. The body of the bone is formed before birth in a procedure called *diaphysis*, which literally means "grow through." This is the first, or primary, center of bone building. Once the bone is

Physical Stress on Bones

A German orthopedic surgeon, Julius Wolff (1836–1902), developed a law that describes the physical stress on bones. Wolff's law states that bones are sensitive to physical demands and adjust their shape accordingly. For example, athletes and manual laborers develop very dense skeletons, and ballet dancers develop slightly enlarged toes.

Space flights might show the reverse to be true: If bones are not used, they might deteriorate. For example, extended space flights may cause NASA astronauts to lose as much as 20 percent of the mass of their bones. Although they follow special diets and exercise, this deterioration may limit extremely long flights (for example, to Mars) because of potential damage to the skeleton.

formed, internal changes by the osteoclasts alter the bone's shape and thickness.

After birth, secondary centers of ossification are called *epiphyses*, and the cartilage plate at each end is called an **epiphyseal plate.** Eventually the epiphysis meets the diaphysis to form one continuous bony structure (note the x-ray of a child's hand in photo). Maturation involves the replacement of cartilage plates and the union of the epiphysis and diaphysis.

BONE HEALTH AND NUTRITION

When Henri-Louis Duhamel did his famous experiments that showed how bone grows, he also demonstrated that diet can make an important difference in bone health. His work inspired others to explore the idea that what people eat affects their health and well-being. It is important to note that in previous centuries, personal responsibility for health care was an unfamiliar idea. The idea that "you are what you eat" began to take hold very slowly.

The formation and maintenance of strong, healthy bones are critical to an active life style. Because bones support the body, protect internal organs, and allow movement (sitting, stand-

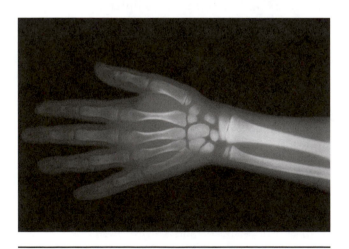

X-ray of a child's hand. © Image Shop/Phototake.

ing, running, twisting, and countless other activities), it is important that they work in the best possible way. Because bones perform silently and are unseen, it is easy to forget the importance of protecting them. Good nutrition does this. A balanced diet includes a variety of foods containing the minerals and vitamins described in the following sections.

Calcium

In the chemistry lab calcium is a white, brittle, chalky-looking element; in the body, compounds of calcium are critical. This element is important for achieving peak bone mass in childhood and for maintaining bone strength throughout life. Calcium strengthens and stiffens bones so that the skeleton can support the body's weight and protect the internal organs from injury. Scientists have found that approximately 70 percent of bone is made of compounds of calcium.

Eating foods with calcium is important not only for the formation of new bone tissue but also for the proper functioning of heart, muscles, and nervous system. Bones are the storage vaults from which the body draws calcium. If there is not enough daily calcium, other systems will withdraw calcium from the bones, which will cause them to become thin and weaken. Weak bones break easily and lead to a loss of mobility because the body cannot repair the damage.

Calcium requirements vary by age and general health status. The minimum daily requirements for three age brackets are as follows:

- ages 10–18: 1,300 milligrams (mg)
- ages 19–50: 1,000 mg
- ages 50 and older: 1,200 mg

Calcium-rich foods include:

- dairy products such as milk, cheese, and yogurt
- dark green, leafy vegetables
- nuts
- fish
- calcium-fortified foods—these have calcium added to them; they include orange juice, milk, dry cereal, and bread

Vitamin D

This vitamin is a partner of calcium and phosphorus. It improves the rate of calcium and phosphorus absorption from the intestine, increasing their amounts in the bloodstream and stimulating action of the parathyroid hormone (PTH), which induces absorption. Vitamin D retrieves calcium from the kidneys before it passes out of the body in urine. Although the required daily allowance (RDA) for this vitamin has not been established, an adequate intake (AI) has been determined. An AI must be sufficient to maintain healthy blood levels of an active form of vitamin D.

The AI for the following age groups has been established:

- ages 19–50: 200 International Units (IU)
- ages 51–69: 400 IU
- ages 70 and older: 600 IU

Only a few foods have natural vitamin D, so some foods are fortified (the vitamin is added to it during processing). It is essential to read the labels on foods to determine if they are fortified with specific vitamins or minerals. The main sources of vitamin D in the diet are as follows:

- fatty fish and fish oils such as salmon, mackerel, and cod liver oil
- other natural sources—egg yolks, liver
- fortified food—milk, margarine, cereal grain bars, dry breakfast cereals

The sun's ultraviolet rays hitting the skin convert a form of cholesterol in the body to vitamin D. Sometimes a vitamin supplement may be recommended if a person is unable to get adequate amounts through the diet.

Phosphorus

Working with calcium to strengthen and maintain bones and teeth, phosphorus aids in the conversion of food into energy. A well-balanced diet includes foods rich in phosphorus, which is found in many of the same foods as magnesium and other vitamins. The following foods are rich in phosphorus:

- dairy products
- egg yolks
- meat, fish, poultry
- beans
- cereals
- nuts
- soft drinks

However, too much phosphorus can interfere with calcium levels.

Magnesium

This mineral stimulates the production of calcitonin, a hormone that raises calcium levels in the bones and prevents calcium from being absorbed into other parts of the body. Magnesium also reduces acid levels in the blood. Bone loss occurs more severely when the blood is more acidic.

The RDA for magnesium ranges from 50 to 400 milligrams. Age, gender, and health determine the appropriate amount.

Foods rich in magnesium include:

- green, leafy vegetables
- seafood
- beans
- tofu
- whole grains—brown rice, millet, corn, oats, barley, wheat
- nuts

Potassium

Like magnesium, potassium lowers blood acidity and slows calcium loss to increase bone formation. It also helps to regulate the heartbeat and to conduct nerve impulses. Potassium is found in most foods, so deficiencies are uncommon:

- leafy, green vegetables
- legumes such as peanuts
- oranges and bananas
- potatoes
- grains and cereals
- meat, fish, poultry
- milk and yogurt

Vitamin C

This vitamin contributes to the growth and maintenance of healthy bones as well as teeth, gums, ligaments, and blood vessels. The recommended RDA is 1,000 milligrams, depending on age, gender, and health. Vitamin C–rich foods include:

- citrus fruit
- strawberries

- cantaloupe
- red and green peppers
- cabbage
- broccoli
- tomatoes

Charts of the amounts of nutrients in most foods are available in nutrition books and many diet books. Also available are supplements that provide the minimum RDA of vitamins and minerals recommended by the American Dietetic Association (ADA).

Because replenishing of bone is necessary, good nutrition affects the skeleton. Inadequate levels of vitamins and minerals can hinder growth and repair. For bone to be constantly renewed, the diet must include proper levels of critical nutrients.

The Embryology of the Skeletal System

Throughout history, people could only imagine what happens before birth. In the eleventh century Trotula, an instructor and physician at the medical school at Salerno, Italy, wrote a treatise called *The Diseases of Women* that included information about pregnancy. The illustrations showed the child before birth like a miniature adult floating in water. Later, in 1604, Italian doctor Hieronymus Fabricius (c. 1533–1609) wrote one of the first books on the formation of the fetus, explaining how nature provides for growth, nourishment, and birth. The same childlike drawings explained the text.

Genetics and embryology were late developing as sciences. It was in the early twentieth century that they began to expand. Development of powerful microscopes enabled the growth (see "The Origins of Embryology").

Traditionally, geneticists have studied **genes** and **chromosomes** and embryologists have looked at the development of fetuses. The body's genes, called the human **genome**, consist of 23 pairs of chromosomes with 30,000 to 50,000 genes. Genes are comprised of deoxyribonucleic acid (**DNA**) molecules that go through complex protein reactions to determine the traits that each individual will possess.

Some areas of current research combine genetics and embryology. For example, to understand why an individual develops a birth defect such as a cleft palate (see Chapter 18), scientists study the interplay between genetics and environmental factors of the uterus. Causes of developmental abnormalities in humans become understandable when looking at embryonic development in the uterus. The human embryo is exceedingly vulnerable

The Origins of Embryology

The study of *embryology*, the science that deals with the formation and development of the embryo and fetus (see Figure 6.1), can be traced back to the ancient Greek philosophers. The field was originally called *generation*. Ancients believed that organisms could arise sexually or asexually (without sperm or egg) or by spontaneous generation (development that had no obvious cause). As early as the sixth century BCE, Greek physicians found that they could study a developing embryo by opening up the eggs of chickens, a technique still used in science classes today.

The Greek philosopher Aristotle (384–322 BCE) described two important models of generation known as *preformation* and *epigenesis*. According to preformation proponents, a complete miniature person existed in the mother's egg or father's sperm and began to grow when the conditions were right. Aristotle thought the mother contributed only a mass of material and that the father provided the true form of the new being.

Centuries later, the English scientist William Harvey (1578–1657), discoverer of the circulation of blood, and his Italian teacher, Fabricius (1533–1619), began to doubt Aristotle's ideas of preformation. Fabrici wrote a work, *On the Formed Fetus*, that attacked these beliefs. While the study of embryology lagged behind other developments in the following years, in the nineteenth century the German researcher Karl Ernst von Baer (1792–1876) discovered the mammalian **ovum**. The field of experimental embryology was begun by the French researcher William Roux (1850–1924) at the beginning of the twentieth century.

to drugs, viruses, and radiation during the first several months when critical systems are forming.

Important areas of investigation include what happens to cells and why they function as they do. The mechanics of the order of how development occurs is called *embryonic induction*. In this process one group of embryonic cells causes an adjacent group of embryonic cells to specialize. The term **differentiation** describes the process during embryonic development when cells take on special structures and functions. Another important area of investigation is *biochemical triggers*: how genes are programmed to cause cells to develop into specific structures, such as arm buds.

Molecular embryologists also note that human development does not stop at birth. For example, bones still grow together and teeth continue to develop. In a modern sense, embryology is in fact the study of the life history of the person. Modern embryology seeks to understand development mechanisms in molecular and chemical terms, investigating at the level of the gene what controls differentiation of specific tissues.

The origin of an individual from fertilization, or the joining of egg and sperm, has implications for orthopedics. Many bone diseases are of genetic origin. For example, brittle bone disease, Marfan syndrome, and vitamin D–resistant rickets can be determined by looking at genes (see Chapter 18). A simple change, or **mutation**, of just one gene may cause a serious genetic defect.

Gene therapy is a treatment for curing genetic diseases by introducing normal genes into patients to overcome the effects of the defective gene. If a defective gene, such as the one that causes Marfan syndrome, could be replaced with a healthy gene, doctors could cure the disorder. Knowledge of genes, chromosomes, and development is essential to understanding orthopedic disorders.

Certain agents, or drugs, may have a **teratogenic** effect on the embryo. The word *teratogenic* literally means "giving rise to monsters," from the Greek word *teratos*, meaning "monster." Pregnant women are advised by their physicians not to take certain drugs unless the situation is life threatening. One of the most unusual medical disasters in history involved the use of a drug called *thalidomide*. The drug was given to pregnant women to combat morning sickness. When their babies were

Figure 6.1. Eight-week embryo.

born, many of them had extreme deformities. The drug had caused birth defects according to the embryonic stage of the mothers' pregnancy. (See "Thalidomide: A Medical Disaster" and photo in color section.)

Scientists have come a long way in understanding the development of the embryo *in utero*, the Latin term for "in the uterus." Such knowledge not only assists in comprehending genetic and birth defects, but it has also formed the basis of the hottest topic in science, **stem cell** research (see Chapter 13). In the early stages of development the **zygote**, or fertilized egg cell, undergoes division to form a blastocyst. Some of these cells eventually develop three layers that become the cells of the embryo. The cells are called *primary germ layers* or *stem cells*; these layers change into cells that will become various parts of the body (see Figure 6.2).

Three embryonic layers make up the body tissue:

- *Ectoderm*—the outer layer that gives rise to the outer epithelium such as hair, nails, and skin; the sense organs; and the brain and spinal cord. Epithelium forms the epidermis and the other structures.

- *Mesoderm*—the middle layer that gives rise to the bones, muscle, connective tissue, circulatory system, and most of the excretory and reproductive systems.

Thalidomide: A Medical Disaster

In 1956, a baby was born with flipper-like arms and other birth defects because the mother had taken thalidomide during her pregnancy. The drug thalidomide had been developed in 1953, but the company stopped its production and gave the rights to the German drug company, Chemie Gruenenthal, in 1957. In Europe in the next few years there was a dramatic increase in birth defects that included *phocomelia*, a shortened-limb defect in which the hands, feet, or both appear like small flippers. The thalidomide exposure blocked the middle stem cell lay or the mesenchyme development in the intermediate limb bud during fetal development. The entire limb bud then formed only the most distal elements at the finger end. This defect was one of the numerous fetal defects caused by the drug. Trunks lacking arms and limbs or both, deafness, cleft palate, and malformed internal organs are only a few of the deformities of twelve thousand children in forty-six countries that can be traced to thalidomide use by their mothers during pregnancy.

In the United States, Frances Kelsey, M.D., had taken a job in 1960 as drug evaluator for the Food and Drug Administration (FDA). The German company wanted to market thalidomide in the United States, but Kelsey thought more tests needed to be done on the compound. Although pressure to approve the drug was great, the FDA stalled. An Australian researcher made the connection between thalidomide and the birth defects. The disaster had one positive effect: It led to stricter procedures for developing and approving drugs. Today, the United States has one of the strictest procedures for approval of new drugs in the world.

- *Endoderm*—the inner layer that gives rise to the epithelial linings, those cells that form the linings of the body cavity, such as digestive tract, most of the respiratory tract, urinary bladder, liver, pancreas, and some endocrine glands.

Of importance to the skeletal system is the mesodermal layer. In the early embryo, this middle layer and certain parts of the outer layer differentiate into bone marrow, bones, tendons, ligaments, fat, and other connective tissue. Stem cell research is based on this point of embryonic differentiation. Because the embryo must be destroyed to study stem cells, questions of ethics hamper this research. Scientists have found stem cells in adult bone marrow, and the possibility of therapy for a multitude of conditions using these cells is under investigation.

DEVELOPMENT OF THE EMBRYO

During the first week, the fertilized egg in the Fallopian tube reaches the uterus. First called a *zygote* and then a *blastocyst*, the contents divide in-

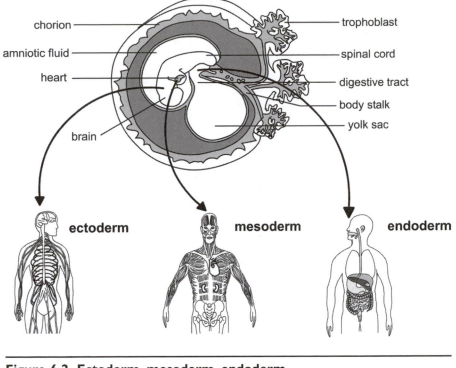

Figure 6.2. Ectoderm, mesoderm, endoderm.

side the confines of the egg but do not increase in size. Around twelve days after conception, a band called the **primitive streak** appears along the back of the mass. This temporary structure lengthens to form an axis that other structures will organize around as they develop. This structure is temporary.

Vertebral Column

At around twenty-one days, the primitive streak gives rise to another temporary structure, the **notochord**, that will form the central axis for the segments of the vertebral column. The notochord is derived from the endoderm and consists of a rod of cells on the underside of the neural tube, a primitive hollow that will eventually develop into brain and spinal cord, or central nervous system (CNS). The notochord induces overlying ectoderm to specialize into the cells that will develop into the CNS.

On each side of the notochord is a column of mesoderm that becomes subdivided into cube-like segments called *primitive segments*, or *somites*. These tiny segments are separated from one another by bits of material arranged in pairs on each side of the neural tube and notochord. To every segment, a spinal nerve is attached. Eventually, these will become the thirty-one pairs of spinal nerves that extend control to every part of the body.

Each primitive segment at first contains a central cavity, or *myocoel*, but the hollow space is soon filled with a core of cells. Around the middle of the fourth week, these cells differentiate into three groups:

- *Cutis*—plate, or dermatome, located on the lateral and **dorsal** side of the myocoel that will develop into skin. *Cutin* is the Latin word meaning "skin," and *derm* is the Greek term.

- *Muscle*—plate, or myotome, situated on the inside of the cutis, it will become the muscles of the segment.

- *Sclerotome*—cells that will form the core of the myocoel and are next to the notochord. The individual sclerotomes begin to fuse from front to back, forming a continuous line along the side and below the neural tube. The cells of the sclerotome multiply rapidly and extend to surround the notochord and at the same time grow forward to cover the neural tube. Now a continuous sheath of mesoderm envelops the notochord and neural tube.

In between the two, a rudimentary form of cartilage is laid down (see Chapter 5).

Also in the fourth week, two cartilage centers make their appearance, one on each side of the notochord. These centers extend around the notochord and form the body of the cartilaginous vertebrae. A second pair grows to become the cartilage vertebral arch. By the end of the fourth week, the two halves of the arch are joined on the back side of the neural tube. The spine is developed from the junction where the two halves of the growing side close. If this closing around the neural tube does not occur at about day 28, a deformity known as *neural tube defect* results (see Chapter 18). In one neural tube defect, spina bifida, an area of the spine is open and parts of the spinal cord protrude.

The portions of the notochord surrounded by the bodies of the vertebrae wither away and ultimately disappear. However, some cartilage structures enlarge and persist throughout life as fibrous cartilage.

Ribs and Sternum

Ribs are formed from the ventral, or costal, processes of the primitive vertebral bows, structures that extend between the muscle plates. **Costal** refers to the structures of breathing. Ribs first appear as tiny bars of cartilage that grow to form a series of arches called the *primitive costal arches*. These bars then join on each side around a middle line. In the cervical and lumbar regions the ends do not meet, and the rib cage develops only where there are thoracic vertebrae. In the sacral region, the segments join to form the lateral parts of the sacrum.

The ventral ends of the ribs are united to one another by a horizontal bar called the *sternum plate*. Opposite the first seven pairs of ribs, these sternum plates fuse in the middle to form the main body of the sternum (or breastbone), the *manubrium*. The xiphoid process, located at the tip of the sternum, is formed by a backward extension of the sternal plates.

Skull

In the beginning, skull development is similar to that of the vertebral column; but as the brain and sense organs expand, the skull develops. The cephalic, or head, end of the notochord becomes partly surrounded by mesoderm. The back part of this corresponds with part of the occipital bone. The mesoderm encases the brain vesicles and is now called the *membranous cranium*. From this inner layer, bone of the skull and the brain are formed. From the outer layer, muscles, blood vessels, and skin of the scalp are formed. In animals like the shark and other fishes that have no bones, the membranes become a complete skull of cartilage. However, in mammals, cartilage precedes the bones that begin to develop at the base of the skull.

The branchial area is located in the region of the throat. Indentations in the ectoderm form the branchial grooves. These grooves separate into a series of rounded bars or arches in which the mesoderm thickens. The dorsal ends of the arches are attached to the side of the head, whereas the ventral ends meet in the middle of the neck. Six arches appear: The first is the mandibular arch; the second, the hyoid; the rest have no names. The mandibular arch will form the lower lip, the mandible, muscles for chewing, and part of the tongue. From the back ends of the arch a triangular process, the maxillary process, grows forward on each side to form cheeks and lateral parts of the upper lip. At about five weeks, the general form of the human body becomes evident (see Figure 6.3).

Nose and Face

During the third week, two areas of thickened ectoderm appear under the forebrain. Olfactory pits (which will be used in smell) form on the side, and two nasal processes form in front. By about eight weeks, the head and nose are formed.

From the third to the fifth month, the nares, or nose, is filled with skin like epithelial

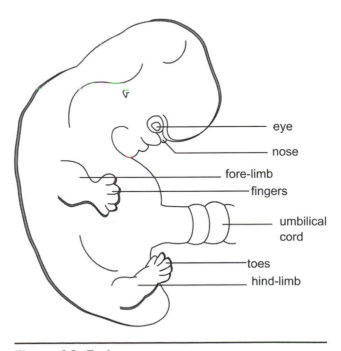

eye
nose
fore-limb
fingers
umbilical cord
toes
hind-limb

Figure 6.3. Embryo.

tissue that disappears to form permanent openings in the nose. The union of parts that form the hard palate begins in the eighth week and is completed by the ninth. The soft palate is completed by the eleventh week. If the areas forming the palate do not close, a deformity known as *cleft palate* will occur (see Chapter 18).

Limbs

In the third week, the limbs begin to make their appearance as small bumps, or buds, at the side of the trunk. From the muscle and skin plates, several segments that extend into each bud push the first divisions of the spinal nerves. Part of the mesoderm of the bud condenses and becomes the cartilaginous skeleton. When these ossify, the bones of the limbs are formed. By the sixth week, furrows mark the three divisions of each limb: the upper arm, forearm, and hand; and the thigh, lower leg, and foot. By week 7, ten finger rays appear and continue to differentiate until weeks 12 through 13, when hands appear. At first, the buds are parallel to the trunk. As the limbs develop, they undergo a rotation, or torsion, through a 90 degree angle to come to the position they will occupy at birth. The upper limbs move outward and forward.

During this initial twelve weeks, the body's solid framework begins to form. Each cartilage precursor that will become bone then turns to bone. However, the skull—which formed through the different process of membrane development—does not ossify. By week 12, the primary of centers of ossification in the diaphyses of most bones have appeared. The secondary centers of ossification are not present until after birth (see Chapter 5).

See Table 6.1 for a complete overview of the fetal development of the skeletal system.

EMBRYONIC DEVELOPMENT AND EVOLUTION

In a human embryo of about six weeks' development, there is a prominent tail. The tail eventually disappears with its only remnant being the coccyx, or tailbone. The term *ontogeny* refers to the period from fertilization though embryonic and fetal development. The term *phylogeny* refers to the development of a group of organisms from the beginning of life to the present.

Several people have noted the many similarities between human development and the development of various species throughout evolution. For example, a simple animal—the amphioxis—has only a notochord. Often during the first eight weeks, it is hard to distinguish the embryos of a frog, mouse, elephant, cat, and human. This led the German scientist Ernst Haeckel (1834–1915) to develop the *biogenetic law of recapitulation*. The law says that ontogeny—development of the individual—replicates the

TABLE 6.1. The Developing Skeletal System

1st week	Fertilized egg in the Fallopian tube reaches the uterus but does not increase in size; by end of first week, the blastocyst implants in the uterus.
2nd week	Primitive streak appears—day 12.
3rd week	Around 21 days the notochord forms the central axis that will become the vertebral column; somites form on each side of the neural groove.
4th week	A period of spectacular growth and differentiation: Somites differentiate; the part called the *sclerotoma* forms the vertebral column; neural tube closes; tail, or caudal, extremity projects.
5th week	First traces of the hand appear; feet, clavicle, and lower jaw begin; primitive palate appears.
6th week	Vertebral column, primitive cranium, and ribs appear as cartilage; furrows mark the three divisions of the limbs; general form of the human skeleton is evident; germs of teeth appear.
7th week	Points of ossification of the ribs, scapula, shaft of humerus, femur, tibia, palate, and upper jaw appear; ten finger rays appear.
8th week	Distinction of arm and forearm occurs; all vertebral bodies are cartilaginous; points of ossification for ulna, radius, fibula, and ilium begin; two halves of hard palate unite; head and nose are formed. The embryo is now referred to as a *fetus*.
9th week	Arches of the vertebrae, the frontal bones, malar bones, shafts of the metatarsal bones, and all the phalanges appear. Union of the hard palate is complete; hands appear.
3rd month	Tail disappears; cartilage arches on dorsal region of the spine close; primary centers of ossification in the diaphyses of most bones have appeared.
4th month	Closure of the cartilage arches of the spine occurs; bone points for the first sacral vertebra and pubis appear.
5th month	Continued development of points of ossification of all bones takes place; some of the tiny bones of the ear develop; ossification of germs of teeth occurs.
6th month	Ossification of part of the seventh and second cervical vertebrae occurs; sacro-vertebral angle forms.
7th, 8th months	Ossification of specific bone parts continues.
9th month	While final points of primary ossification continue, the fetus gains weight and normally assumes the birth position of head first.

short and rapid evolution of a group of organisms, or phylogeny. He thought the similarity occurred because of a common descent from a more primitive form. Today, scientists think Haeckel was wrong, because at no time do humans replicate forms to which they may have been related.

The process of development before birth is complex and mysterious. Although by now scientists have a fairly good idea of what happens at each stage, their investigations have shifted to finding out how and why. Combining embryology with genetics and the study of proteins not only will help find answers to combating birth defects and genetic diseases but also will lead to a better understanding of life itself.

The Ancient World: Bones and the Beginnings of Orthopedics

One of the greatest thrills that can be experienced is the discovery of something no one has ever seen. Columbus (1451–1506), Vasco da Gama (1460–1524), and other adventurers found new and exciting worlds. But a group of medical pioneers were involved in discovery of a different kind: They explored the frontiers of the skeletal system (see illustration). Some of the pioneers learned about the system itself, and others were involved in the search for cures and healing.

Some of these pioneers' ideas may sound strange today, but they all contributed knowledge that enables us to know about the skeletal system and how it works. A philosopher who added much to scientific thought, Karl Popper (1902–1994), said that the advance of knowledge consists mainly of modifying earlier knowledge. And so it is with current knowledge of the anatomy and physiology of the skeletal system and of the art of orthopedics (the branch of medicine that treats conditions of the bones). In fact, our knowledge today, which is still evolving, began with ancient—even prehistoric—practices.

In ancient times, battles between warriors inflicted great wounds. The warriors knew the wounds must be closed, so they created an ingenious suture. As one person held the edges of a wound tightly together, another would put ants or termites on the skin, and the powerful pincer jaws would attach to the flesh. Once securely in place, the body of the ant or termite would be severed, allowing the jaws to remain firmly in place to hold the wound together.

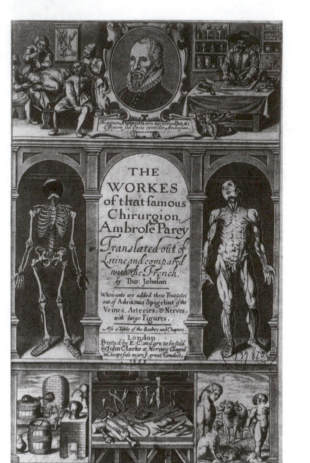

"The Works of that famous Chirurgion Ambrose Parey,"
1510?–1590. © National Library of Medicine.

This procedure was one of several techniques to close wounds. Deposits from the cultural Paleolithic period of the early Stone Age, have yielded bone needles with an eye hollowed out for a threadlike material to pass through. Most likely these were used for suturing, or stitching up, wounds. Early humans probably also used crude splints and performed amputations of limbs or fingers. In fact, skeletal remains from the many cultures indicate that surgery did take place. One of the most interesting procedures in many cultures was trephining, the drilling of holes in the skull to release "trapped evil spirits" (see illustration).

Early humans were aware of the importance of bones in their struggle for existence. Civilizations soon developed that diagnosed and treated bone problems.

MEDICINE IN CIVILIZATIONS OF ANTIQUITY

Mesopotamia: Healing Practices

The Fertile Crescent, located in the area of the Tigris-Euphrates River, is called the *cradle of civilization*. It was in this warm and fertile region of the Middle East, located now in the country of Iraq, that the world's first great civilizations appeared around 3000 BCE.

In the wedge-shaped characters of cuneiform writing preserved on baked clay tablets from the ancient cultures of Babylon and Assyria are treatises on medicine with diagnoses and remedies. The chief text, called "The Treatise of Medical Diagnosis and Prognosis," recognized drugs and surgery as constituting one of three types of healing. The famous king of Babylon, Hammurabi (1717–1686 BCE), engraved a legal code on a 6.5-foot-high, stele, or pole (now displayed in the Louvre in Paris) that included medical instructions for physicians. If the physician successfully performed a major surgery on a lord with a bronze spear, or lancet, he would get 10 shekels (gold or silver coins), but if he caused the lord's death, his hand would be chopped off.

The medical practices of these groups might be considered systematic sorcery. Healing practices involved prayers, magic, and sacrifices, as well as surgery and drug treatments.

Egypt: Medical Texts and Social Role of Bones

In the third millennium BCE, Egypt rose to power under the pharaohs. The earliest written record from the great pyramids

A surgeon bores a hole in a patient's skull. From Croce, Giovanni Andrea del, *Chirurgurgiae . . . libri septem*, 1573. © National Library of Medicine.

in the Valley of the Kings dates from about 2000 BCE. Mummified bodies, wall paintings, and hieroglyphics show that people then suffered some of the same bone problems that exist today. Paintings reveal some of the orthopedic practices of the time. Mummies had splints that were made of bamboo, reed, wood, or bark, padded with linen. The earliest record of the use of a crutch is in a carving on the portal of a tomb made in 2830 BCE.

A British Egyptologist exploring a tomb near Luxor in 1862 found a papyrus (a scroll made of reeds) that was later named after him: the Edwin Smith papyrus. The text is believed to have been written by the physician Imhotep around 1600 BCE and is sometimes called *a book of wounds*. This book told how to treat a dislocation of the mandible: If the mouth is open and cannot close, the physician must put two thumbs upon the end of the two rami of the mandible inside the mouth and the fingers under the chin and *force* the parts back to their normal position. Also mentioned were signs of spinal injuries and treatment strategies for fractured clavicles and other injuries.

The papyrus also referred to dressings, adhesive plasters, braces, plugs, cleansers, and cauteries (instruments for burning to destroy dead tissue). Broken bones were set in ox-bone splints, supported by resin-soaked bandages. The document also described the examination of the limbs and showed pulses as being blood vessels that go to the limbs. Injuries were classified into three groups: a condition that could be treated, a problem that could be tolerated, and a condition or problem that could not be treated.

Another document—the Georg Ebers papyrus of ca. 1550 BCE, from Thebes—is probably the oldest surviving medical book dealing with a number of conditions and treatments. Surgery was limited to the repair of injuries and bone fractures. Sutures and cauteries were used, and wound dressings to promote healing called for honey with grease or resin. No surgical instruments were used. Knowledge of anatomy was limited to bones and major organs. The Egyptians did not have taboos against touching the dead.

The physical aspects of sports and games were important. Running, swimming, rowing, archery, and wrestling were popular in the cities of Thebes and Gizeh by the Nile. Hieroglyphics show boxing as a common sport. The first medical text for sports was the Kahun papyrus that originated around 1900 BCE.

Combining the mystical with medicine, the Egyptians—like many ancient cultures—gave symbolic meaning to bones. In the first century CE, the Greek historian Plutarch (ca. 46 CE–after 119) told how the Egyptians would "invite," a skeleton to their festivities. The skeleton would be seated among the living to remind them that death makes everyone alike. The bones admonished partygoers to take the opportunity to reflect on death and the proper way people should live. Early on, the skeleton became a reminder of death.

In addition to spiritual implications, sometimes the bones played a part in social practices. Pharaoh Akhenaton's (1379–1362 BCE) skull was naturally elongated at birth. In paintings his queen, Nefertiti, and their daughters share the same profile. This royal deformity set a fashion for artificially lengthened skulls, accomplished by binding the head during early years, that swept throughout the Mediterranean and Africa. Skull shaping was also widespread among Indians of the Americas.

India: Early Hindu Medicine

In India, medicine developed in the Brahman period from 800 BCE to 1000 CE. The famous medical textbook *Susruta Sambitu* has a large section on surgery that describes over 101 instruments, including the one deemed most important—the hand. This book describes suturing, cauterizing, and excising. One specific procedure for plastic surgery of the nose is called *rhinoplasty*. As punishment for adultery, the society demanded cutting off the noses of the guilty parties. The nose could only be restored by rhinoplasty. Susrata, a famous surgeon, gave instructions predominantly for anatomical dissection and improving surgical techniques. Another great Hindu, Caraka, who lived during the millennium before Christ, is regarded as the father of Indian medicine.

China: *Yin* and *Yang*

Chinese medicine developed along different traditions from earlier civilizations. The classic text, *Huangdi Neijing*, represents medical teaching at the time of the famous Yellow Emperor, Huangdi (2698–2598 BCE). The Chinese believed in a life force called *qi* and a complex system of *yin* and *yang* (complementary forces, feminine and masculine characteristics). Blockages, or disturbance of *qi*, produced symptoms of illness. Acupuncture stimulated the flow of *qi* by means of needles inserted at appropriate points.

The Chinese had a unique way of modifying the skeleton. For over a thousand years, women of high class had their feet bound as infants. Called the *lotus foot*, the procedure involved binding the heel of the 5- or 6-year-old as closely as possible to the forefoot, excluding the big toe. As she got older the bonds got

A seated Chinese woman with bound feet, ca. 1900. Courtesy of the Library of Congress.

tighter, stunting the natural growth. The results were a steeply curved instep with a deep cleft below the arch. The lotus foot was prized for its erotic appeal (see photo).

The Americas: Herbs and Anatomy

The ancient civilizations of the Mayas, Aztecs, and Incas had a religious and magical aspect to their medicine that incorporated many herbs, some of which have become important medicines today. The Aztecs developed their knowledge of anatomy through the practice of human sacrifice. Skull shaping was practiced among many tribes. The Incas of Peru, like the ancient Egyptians, mummified their dead.

In North America, the Chinook tribe of the Pacific Northwest would stretch an infant on a board and fix another board over its forehead. After a year, the child had a flattened head, broadened crown, and receding forehead.

The Hebrew Tradition: Bones as Symbols

The Hebrews considered bones an important part of their history. In the book of Genesis, God caused Adam to fall into a deep sleep, took a rib from him, and made the woman Eve. Several biblical passages tell of bones as a source of renewed life. One of the most graphic is the vision of Ezekiel and the valley of the dry bones. The Lord sat Ezekiel down in the midst of a valley strewn with dry bones and made him watch as the bones reassembled and flesh and skin appeared, symbolizing what would happen to the children of Israel. Another Hebrew tradition claims the source of life is in the *luz*, a bone that is not specifically described but might be the sacrum. This bone is considered to be like a seed from which a resurrected body can arise.

GREECE AND ROME

Ancient Greece: Early Orthopedics and Philosophy

By 1000 BCE, communities around the Aegean Sea developed a civilization that emphasized soundness of mind, body, and soul. The Olympic Games in 776 BCE emphasized sports medicine. Athletes started training at gymnasiums under the supervision of sports physician trainers, or gymnates. These trainers developed a routine of physical exercises, massage, diet, and bathing, as well as minor surgery. Throughout Greek civilization, ideals of manliness required keeping one's physique in peak condition, as evidenced in Greek painting and sculpture. Homer (eighth century BCE), the Greek epic writer, gave glimpses of early Greek medicine. In the *Iliad* he told of 147 cases of battle wounds and related how the physician to King Menelaus of Sparta extracted an arrow, sucked out the blood, and applied a salve. The *Iliad* also has references to deformities.

Homer's writing revealed how Greek society drew heavily on sacred healing. Apollo was the god of healing. Asclepius, the son of Apollo and a mortal, became the god of medicine. (Sometimes the name is given as *Aesculapius*; the two are the same.) Asclepius is usually portrayed with a beard, staff, and snake. His staff, the **caduceus**, has two snakes intertwined like a double helix on a winged staff; shedding of a snake's skin symbolized the renewal of life. The symbol is the insignia of the modern medical profession (see "The Symbol of Medicine"). The god is often shown with his daughters: Hygeia, or "health and hygiene"; and Panacea, or "cure-all."

Little is known of Greek medicine before the appearance of written texts in the fifth century BCE. In the period between 430 and 330 BCE, a very important Greek text—the *Corpus Hippocrates*—was collated. Hippocrates was born on the island of Cos in 460 BCE and died in 370 BCE. Credited with bringing a systematic and scientific approach to medicine, he was the first to define the role of a doctor in society. Centuries later, graduates of medical schools still take the Hippocratic Oath, which is based on this ancient physician's ethical treatment of patients.

Several volumes of the text are related to orthopedics. Hippocrates had a thorough understanding of fractures, which he treated with bandages and splints. Occasionally clay or starch was used to strengthen the bandage,

The Symbol of Medicine

A person walking into a hospital, a physician's office, or a pharmaceutical company might see a stylized version of the symbol of the physician: the staff of Aesculapius, or the *caduceus*. The modern version is different from the original in that it has wings and twin snakes.

The first caduceus represented the magical rod of the Greek god of healing, Apollo, who gave the rod to Hermes (same as the Roman god Mercury). According to Greek mythology, the symbol arose when Hermes delivered the infant Aesculapius as his dead mother lay on her funeral pyre. Aesculapius was then trained by Chirion, the healing centaur. Aesculapius's staff was a rough-hewn cypress branch encircled by a snake, two very common objects in the Greek Isles. The thick piece of cypress represented strength, and the snake represented wisdom. Aesculapius is mentioned by the ancient poet Homer for his miraculous healing acts on the battlefield.

Aesculapius became the god of medicine and his descendants formed the cult of Asklepia, which spread throughout the Greek Isles and Mediterranean Sea. By 400 BCE, over three hundred temples were dedicated to healing. Hippocrates was educated as a member of this healing cult.

The wings—symbol of Mercury, the messenger to the gods—were probably added by the Romans. Mercury used the caduceus as a symbol of peace. One myth tells how Mercury saw two snakes fighting and threw the rod between them. They stopped their battle and wrapped themselves around the rod.

Throughout history the two symbols were used interchangeably until 1500. At that time, a printer in Italy decided to put the symbol with wings on the medical books he printed.

The symbol's use was solidified in 1902, when a U.S. army captain decided the army medical corps needed more than a cross as its symbol and encouraged the force to adopt the caduceus with wings. The symbol then became entrenched, although sometimes one still sees the rod without the wings.

---

making a cast. He knew the principles of traction and counter-traction and developed special splints for fractures of the tibia. He also developed the *scamnum*, or Hippocratic bench, for traction. Made of wood and iron, it mechanically stretched and levered forces for the physician. The device had crankshafts and levers laced with straps and cords (see illustration). (In an unintended variation of its use, medieval Europeans made the scamnum a rack for torture.) Another Hippocratic procedure was to hang the patient's body upside down, using body weight to create the forces necessary to realign the spine.

One volume described the joints and how to treat dislocations of the shoulder, collarbone, mandible, knee, hip, and elbow. The volume also explained how to correct clubfoot and how to treat a compound fracture with

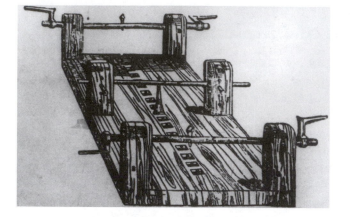

The Scamnum of Hippocrates. From a woodcut in Guido Guidi's *Chirurgia*, 1554. © National Library of Medicine.

pitch and wine compresses. Bandaging was not to be forced. Probing into a compound fracture was to be avoided.

Greek medicine was soon mixed with philosophy. Practicing health became a great philosophical debate. Studying Hippocrates, the philosopher Plato (428–347 BCE) developed a series of analogies to divide human nature into three functions—reason, spirit, and appetites—all of which were considered to be in conflict with each other. In his treatise *Timaeus*, he linked medicine with philosophy, health, and politics.

Plato's pupil, Aristotle, launched a program of empirical investigation emphasizing observation and proof. Some people call him the father of the scientific method. For two thousand years, his ideas formed the framework for scientific investigation. Because of the military conquests of his pupil, Alexander the Great (356–323 BCE), his thoughts and beliefs were spread to the then-known world. His study of anatomy was taken up and revised by the next generation in the work of Diocles of Carystos (320 BCE) and two physicians working in Alexander's city, Alexandria, in Egypt.

In Alexandria, Herophilus of Calcedon (330–260 BCE) is believed to have practiced human dissection in public. He first divided nerves into sensory and motor components. His contemporary, Erasistratus of Chios (330–255 BCE), reportedly dissected "inferior" living human beings, such as criminals and slaves. Around 100 BCE, Hegator described in detail the anatomical relations of the hip joint and recorded a description of the ligaments.

Ancient and Medieval Rome: Surgery and Sports Medicine

Additional scientific progress came from Rome. The Romans were great soldiers and needed doctors on the battlefield. They brought care to the wounded and encouraged the building of hospitals and special training for medical officers in the field. (See "Military Medicine in Ancient Greece and Rome.") A field physician, Scribonius Largus, accompanied the Emperor Claudius (10 BCE–54 CE) to Britain in 43 CE.

Aurelius Cornelius Celsus (42 BCE–37 CE) was a nobleman and wealthy landowner in the Middle Ages who sought to compile books that were the best of animal and human science. Using much of Hippocrates's work, com-

Military Medicine in Ancient Greece and Rome

The ancient Greeks and Romans learned about the skeletal system in battle. In fact, Hippocrates (ca. 460–ca. 377 BCE) said, "He who desires to practice surgery must go to war." And wars were regular occurrences for both the Greeks and the Romans. During these times, physicians could find out about human anatomy and try out new surgical tools. The medical care of the wounded was a hallmark of the Greek armies, and their methods heavily influenced the Romans.

Before the Greek influence, the Romans did not have medical services to the wounded. Then, in the second century, soldiers took care of other soldiers. The Emperor Trajan (53–117 CE) implemented a system of *medici*, or doctors. At first, they were not trained but could dress wounds and do simple surgery. During the time of Galen (129–ca. 199), the Roman armies began having trained doctors and set up field hospitals. Much of the information about the Roman practices comes from the writings of Galen, the writings of Celsus (42 BCE–37 CE), the books of Paul of Aegina (ca. 640), and archaeological excavations of Roman fortresses and battlefield sites.

The Roman legions were very efficient and expanded the Greek medical system for their own purposes. At first, the field hospitals made little distinction between medical and veterinary services, so soldiers and their wounded horses were usually kept in the same hospital. Later, the soldiers were cared for by *medici* who were either Greek or Greek-trained. All the physicians had to pass a medical examination. Carriages for their medical supplies and for the wounded were placed in the middle of marching columns.

Celsus was very interested in avoiding battlefield infection, and he noted that the damaging condition was more likely to occur when bone, sinew, cartilage, or muscle was injured. Amputations were performed, but not often. Celsus discussed how to amputate the leg just above the infection, sawing through the bone as close to the good flesh as possible but leaving enough of the flap of skin to cover the remaining bones. Even though much of the finesse of Roman medicine was lost after the fall of the Roman Empire, the revival of classical writing in the eighteenth and nineteenth centuries greatly influenced medical techniques in warfare at that time.

bined with his own experiences, he wrote in elegant Latin. His works became very popular during the revival of classical learning in the Renaissance. One of his works was the first medical text printed in 1478. He added considerable knowledge of the skeleton and gave an account of the limbs and surgical treatment.

Claudius Galen of Pergamon (129–ca. 199) was a famous philosopher born in Greece who wrote hundreds of books. He became a surgeon to the Roman

gladiators and added an experimental dimension to Greek medical literature. Although most of his dissections were of animals such as gorillas,
sheep, pigs, and goats, he occasionally would study the skeleton of a rob-
ber picked clean by buzzards. Because he studied many bodies that were
slashed open in gladiatorial bouts, Galen was able to write in his treatise,
On Bones, that one must not merely read about bones in books but actually
observe and handle each bone. He eventually worked in Rome as a physician to the emperors.

Known as the "father of sports medicine," Galen described how muscles
move the skeleton. He described the cervical ribs and destructive bone diseases such as osteomyelitis. Galen used Greek words—*kyphosis, lordosis,*
and *scoliosis* (see Chapter 16)—to describe conditions of the spine. He also
devised methods for correcting these conditions and classified bones into
those having a hollow marrow cavity and those, like the bones of the skull,
that do not. He recognized twenty-four vertebrae in the spine and gave the
name *coccyx* to the human tailbone. He also described the sacrum, the block
of fused vertebrae at the base of the spine.

Galen expanded upon classical Greek medicine but several other physicians also added to knowledge about the skeletal system. Julius Pollux, a
Roman contemporary of Galen, wrote the book *Onomasticon*, which introduced many anatomical terms such as *trochanter* (the bony knob at the top
of the thigh where muscles are attached). Soranus of Ephesus (ca. 100 CE)
is said to have described the condition known as *rickets* (see Chapter 17).
Another physician, Rufus (70–120), who was trained in Alexandria but
spent time in Rome, investigated *tendon ganglia* and their treatment by *compression*. Antyllus (third century) is said to have practiced *subcutaneous
tenotomy* (cutting of tendons) to relieve contractions around a joint. He used
both linen and catgut sutures for his procedures. Also, various drills, saws,
and chisels were developed during this period.

Overcome by epidemics, natural disasters, and decay from within, Rome
ultimately fell to barbarian hordes that invaded from the north. The decline
and fall of the Roman Empire brought the decline and fall of scholarly
thought and learning, especially in medicine. As the Dark Ages began, learning gathered by the Greek and Roman physicians became regarded as absolute truth.

8

The Rise of Christianity through the Dark Ages: Bones, Religion, and Superstition

During the period from the rise of the Christian Church to about 1450, Europe was plunged into the Dark Ages: centuries of ignorance, superstition, and mental stagnation. Bones took on an aura of superstition and mysticism. Relics of the bones of saints and martyrs became magic objects of healing and adoration.

In 313, the Emperor Constantine turned to Christianity and moved his capital to Constantinople, later Byzantium (now modern Istanbul). The Roman Empire split into East and West. The East became strong, but the West collapsed with the fall of Rome in 476 as hordes of barbarians from the north pillaged and sacked the city.

MEDICINE AND THE CHRISTIAN CHURCH

Medicine and religion were closely related at this time. The naturalistic and empirical, or scientific, medicine of Hippocrates (ca. 460–ca. 377 BCE) and Galen (129–ca. 199 CE) was blended with religion and pagan superstition. The Church taught that a supernatural plan and purpose provided for the salvation of the soul. Christianity emerged from Judaism, whose traditions and sacred writings did not have a place for the professional physician. To the Hebrews, Jahweh—or God—was the only healer. However, the Hebrews did develop teachings about the body and its well-being. They acknowledged the importance of physical and spiritual cleanliness and had rules that governed personal life.

Christians had a more complex attitude about healing and medicine. Luke, the gospel writer, was himself a physician; but healing was portrayed as a matter of faith, involving prayer and laying-on of hands. The Christians adopted many pagan healing practices. Healing miracles were credited to the direct action of saints or their relics. (The term *relic* refers to the bones or other body parts of a saint.) By the fourth century, large cults developed around these objects. Cult leaders even dismantled the bones and chopped them into little pieces so that several shrines could share the parts.

Shrines were raised on pagan temples of healing. Two physician brothers, Damien and Cosmos, replaced the Roman gods Pollux and Castor as the patron saints of medicine. Practicing in Aleppo (modern Syria), Damien and Cosmos were martyred in the fourth century. Visitors to the brothers' tombs reported miraculous cures. A famous legend tells of a man with gangrene in one leg who prayed at the Shrine of Sts. Damien and Cosmos in Rome, and his faith enabled him to survive a miraculous operation. Surgeons removed the bad leg and sewed on the leg of a black Moor (a Spanish Moslem) who had just died. When the man awoke the next morning, he had two sound legs—one white and one black. This extraordinary story of the first graft that preceded modern transplantation by many centuries has been the subject of hundreds of paintings.

Bones of the saints and holy water became the accepted remedies that were part of magical healing. As a result of this mindset, the ancient Greeks' intellectual inquiries in the explanation of ideas was stifled. Looking for the cause and course of disease and disorders had no part in magical healing.

Another force during this period was the rise of Galenism. Indeed, Galen's words became the authority, and a number of physicians made collections of his work. Oribasius (325–397), a native of Pergamum, gathered Galen's writings into four volumes that told how mechanical devices, including screw traction and elaborate multiple-pulley systems, could treat fractures. Thus Galen's ideas about the nature of illness, humors, and medical procedures became entrenched for many centuries to come. Ancient doctors thought the body was composed of four fluids or humors: blood, phlegm, black bile, and yellow bile. An imbalance of these fluids caused disease.

The belief in astrology, the study of how the position of the stars affects human beings, predates the Greeks and had great influence in medicine. Developed by the Babylonians between 664 and 562 BCE, astrology linked the signs of the zodiac with specific parts of the human body. These signs controlled the time for medical treatments such as blood-letting, or bleeding a person from the veins to purge him or her of disease.

Medicine also found a home in monasteries, where friars and monks tended the sick and the lame. Founded by Saint Benedict (480–547), monastic medicine flourished in such places as Monte Cassino in Italy, Tours in France, St. Gall in Germany, and Oxford and Cambridge in England. However, although the Benedictines were committed to caring for the sick and the disabled, they were forbidden to study medicine because they believed that God alone cured all infirmities.

Paul of Aegina

In the eastern part of the former Roman Empire known as Byzantium, a great medical school developed in Alexandria in North Africa. Byzantine medicine made use of the works of earlier Greek doctors, especially Galen. Paul of Aegina (625–690) wrote the *Epitome of Medicine*, seven books based on Hippocratic texts. The sixth book dealt with fractures and dislocations. He described how a thigh is dislocated from its socket and recommended using the hands to set and to apply ligatures and cords. He recognized that the underpart of the nose is cartilage and does not fracture but that it may be crushed, flattened, or distorted, and that the upper bony part can be fractured. He also described how to trephine the skull. Trephining is an ancient practice of cutting holes in the skull (see "Holes in the Head").

Although classical learning declined, medieval scholars in Europe clung to whatever written documents they had. Paul revered the ancients and believed they had written all that could be expressed on the subject. As a disciple of Hippocrates and a dedicated Galenist, he preserved their ideas and teachings and only occasionally added his own. He did introduce a technique for mending shattered kneecaps and was bolder than Hippocrates in treating spinal disorders. Paul's greatest service was in preserving the works of the ancient Greeks that otherwise might have been lost.

Holes in the Head

Many early people believed that evil spirits or demons caused disease and pain. By drilling a hole in the head the demons were released. Trephination is one of the few ancient medical practices that are substantiated by evidence. Holes were drilled with a tool resembling a carpenter's bit with a handle. Headaches, skull fractures, epilepsy, and some forms of mental illness were treated in this way.

The practice apparently was widespread. Residents of neolithic Gaul, Bohemia, North Africa, Tahiti, New Zealand, and Peru used tools and knives of obsidian, bronze, and stone to bore as many as five holes in a given skull. Some patients survived, as evidenced by signs of healing in the skeletal remains of their skulls. The wound in the head was covered with gourd, stone, shells, silver, or gold. In Europe, the round plugs that were cut out of the bone were worn as charms. Trephination lasted well into the Middle Ages.

MEDICINE AND ISLAM

The prophet Mohammed (571–632) was born in Mecca (now in Saudi Arabia) and as a young man had a vision in which Allah, or God, revealed the Koran (the holy book of Islam) to him. He then began to preach that "There is no god but Allah, and Mohammed is his prophet." He developed such a large following that within a century of his death the Islamic Arabs had conquered half of the Byzantine Empire, Asia, Persia, Egypt, North Africa, and southern Spain. The Arabic language became very important in the eastern world just as Latin and Greek were in the western world. Sects of heretical Christians had fled from the western empire and set up universities with medical schools in the east. *Heretics* were those who did not hold to the exact dogma of the Church. The Nestorians, a Christian sect, went to Persia and established a medical school at Gondisapore. In 636 the Arabs conquered this region. But instead of destroying the university, they encouraged it. The Nestorians, Jews, and Arabs all continued to translate the great Greek works into languages of the conquered areas. They added Chinese, Indian, and Persian medicine and thus spread classical medicine through Syria and Persia and to the rest of the Arab world.

In this age of translating old texts, several centers developed. One was in Baghdad, Persia, where hundreds of Greek texts were rendered into accurate and eloquent Arabic. Thus Galen became the father figure of Arabic medicine. The period 750–900 became known as the Age of Translations. Because the Moslem, or Saracen, Empire expanded into Europe and most of the known world, it greatly influenced the medical thinking of the West from the twelfth to the fifteenth centuries.

Great Sages of Islamic Medicine

Rhazes (864–925; his Arabic name was al Razi, or Abu Bakr Muhammed ibn Zakari-ya-Razi) was born in Persia and studied at Baghdad as well as in Palestine, Egypt, and Spain. He was a celebrated teacher and physician in the cities of the Moslem Empire. He wrote the *Mansurian Book of Medicine*, which addressed anatomy, physiology, medicine, surgery, and pathology. Although most of his treatments were an extension of Greek medicine, Rhazes is credited with the use of plaster of Paris in the tenth century. He added water to a substance now known as calcium sulfate to produce a hard crystalline material and used this to coat fractures and other injuries of the limbs. Rhazes built one of the first great hospitals in Baghdad. Determined to find the most healthful location for this hospital, he hung pieces of meat at likely sites. Any site where the meat did not rapidly decay indicated a place where the chances of infection were low. Revered as the greatest physician of the Islamic world, he was the first to compile case histories that gave insight into both doctor and patient.

Avicenna (980–1037; Abu Ali Husayn ibn Abdullah ibn Sina) was physician-in-chief to the hospital at Baghdad and served several important rulers of the area. Known as Prince of Physicians, his contributions to medical writing and his influence on medical thought in the Middle Ages were second only to that of Galen. Like Rhazes, he wrote and organized medical thought but concentrated more on surgery. He expected doctors to master surgical techniques for treating a wide variety of diseases and disorders. Because sports medicine was practiced in the Islamic world to prepare for battles, in his book *Canon of Medicine* Avicenna addressed health and physical and mental preparedness for battle.

Another physician and writer, Albucasis (936–1013; Abul Qasin Khalaf ibn al-Abbas al Zahrani), was born in Cordoba, Spain, the capital of the Islamic west. His most celebrated work deals with surgery. He described operations for cauterization (burning or searing) of wounds, dental practices, setting fractures, and treating dislocations; he included over two hundred illustrations of medical and dental instruments. His plasters incorporated a variety of fine dust from flour mills mixed with egg whites. He also experimented with clay, acacia, myrrh, and aloes for casts. He attacked the **bonesetters** (see Chapter 9) as untrained imposters who went around the countryside breaking bones and resetting them. His surgery text especially influenced the development of medicine in Europe.

Many scholars throughout the ages have criticized the work of these physicians as stifling independent thought. Some of their works were highly critical and even questioned the conclusions of Galen. But in general the physicians still were considered Galenists. Their contribution to medicine was not in originality and novelty but in the thoroughness with which they organized existing knowledge. Over five thousand medical manuscripts in Arabic still can be found in libraries. Without the Arabs, documents that spurred the Renaissance and the great rebirth of medicine would never have existed.

THE SLOW AWAKENING

During the Middle Ages, Salerno, a town in present-day southern Italy, was located at the crossroads of the world, and it became a cultural, economic, and ethnic melting pot. The Benedictines had started a monastery in Salerno and one monk, Alphanus (d. 1065), traveled to Constantinople and brought back Greek texts. He showed these to three other scholars—a Jew, an Arab, and a Greek—and they established in this small town a university that attracted gifted teachers and students. The school soon became secularized and drew away from the Benedictine and Catholic traditions.

During the eleventh and twelfth centuries, there began a return to the classical tradition that eventually led to the Renaissance. Scholars at this time studied anatomy by dissecting animals. Then came a great breakthrough: the courage to dissect human bodies, usually those of executed criminals, even though both the Catholic Church and Islam forbade dissection. The school at Salerno became a vehicle for the slow movement of Greek and Roman medical knowledge into Europe.

Salerno was located in an ideal spot not only for a medical school but also for a health resort. It received balmy breezes from the ocean, had mineral springs, and was close to the mountains for pure, fresh air. Popular health books emanated from this school. One was called the *Saleritan Regime of Health*, a book of verses compiled in the thirteenth century by Arnald of Villanova (1240–1311). This work extolled a healthy life style and gave tips for living from youth to old age, highlighting hygiene, exercise, diet, and temperance. This first home health manual was very popular and was later printed in 240 languages.

Salerno created a ripple effect for learning. From this area in southern Italy, the center of medical studies moved north to Bologna, Italy. William of Salicet (1210–1280) studied anatomy and surgery there, following the works of Hippocrates and Galen. He wrote a treatise, the *Cyurgia*, that reveals his dedication to a thirteenth-century return to Greek medicine. Although he had little new information to add, he was the first European to link *crepitus* (the sound made by the ends of bones rubbing together) to fracture. His knowledge of the neck enabled him to align dislocated vertebrae without using mechanical devices. During this time, the official position of the Catholic Church still forbade dissection, with several edicts from various Popes forbidding it.

The first recorded public human dissection was performed in Bologna around 1315. Mondino de' Luzzi (c. 1270–1326) set the stage for the actual study of the human body. He was born into a medical family and graduated from the school in Bologna. Eventually rising to hold a chair of medicine there, he was one of the first to write a standard text on anatomy built on the personal experience of human dissection. He encouraged people to boil bones before studying them not only for the sake of safety but also because it was considered less sinful. Because he was relying on Galen and Hippocrates, his anatomy book perpetuated many errors. For example, he described a five-lobed liver and a three-chambered heart.

Mondino and his colleagues began to encourage open dissection, using the cadavers of executed criminals, and justified it in terms of natural philosophy. They declared that the body revealed the wisdom of the Creator; they did not mention the benefits of helping man with surgery. Mondino made these dissections a public spectacle of both instruction and entertainment. They usually took place in a church in the cold of winter to slow down deterioration.

By performing dissections, these scholars were committing heresy in the eyes of the Church. In the fourteenth century, Franciscan and Dominican priests and monks led a counterattack against the heretics and intellectual innovators. This led many scholars to flee from Italy to France. Guido Lanfranchi di Milano (1250–1308), defied the counter-reformers and fled to France. He became known as Lanfranc. With these moves, the center of European surgery moved from Italy to France, where it remained for many centuries.

Lanfranc, an Italian, ultimately went to Paris to escape the crackdown of the Church against those teaching dissection. First settling at Lyons, a place friendly to scholarship, he later moved to Paris. He became an eminent surgeon and wrote *Chirugia Magna* in 1266. He stressed the importance of cleanliness in wounds and believed air was a source of problems for wounds.

Guy de Chauliac (1300–1368), a professor at Montpelier in France, wrote two books on surgery (see photo in color section). He described the treatment of femoral fractures with traction by weights and pulleys, and he stressed the importance of having an assistant to hand him all necessary tools. He attacked the **barber-surgeons** (see Chapter 9) who—he said—knew no anatomy and carved the body like the blind carved wood.

Book-driven anatomy was the hallmark of this period, as physicians perpetuated the teachings of Hippocrates and Galen with few new observations. Medieval illustrations were childlike, simple drawings. The most common type of illustration was the "zodiac man," a male figure marked with bloodletting and zodiac signs.

The Late Middle Ages was a period of gloom and despair. Plague and disease attacked peasants, merchants, and princes alike. But changes were on the way. Following the Italians, Spanish doctors performed their first public dissection in 1391; the Viennese followed with a public dissection in 1404. The Italians introduced practical medicine with case histories and a hands-on approach. Physicians were being recognized for their skills and knowledge and soon would be competing with other practitioners such as barber-surgeons and bonesetters. Rebirth, or renaissance, slowly appeared in many places in Europe.

9

The Renaissance: The Skeletal System as Anatomy Text

The term *renaissance* literally means "rebirth"; it designates a period of great upheaval in Europe roughly between 1300 and 1650. The old patterns of economic, political, and intellectual stagnation were giving way to a new age of exploration of the world, the mind, and the human body. Superstition and medical quackery were still part of everyday life. However, the relationship between the cultural renaissance and the medical renaissance was complex. Changes that began in twelfth-century Italy eventually led to the medical renaissance of the sixteenth century.

Two great inventions happened during this time, which was the beginning of the Scientific Revolution: the printing press and gunpowder. Bonesetters (see sidebar), barber-surgeons, and people who practiced herbal medicine, were the caretakers of the majority of people. Inquiries into basic science, the beauty of art, and the horrors of war all combined to increase the knowledge of the skeletal system. By the end of this period, physicians and surgeons had information that no previous generation had possessed. Applying new thought would have been impossible without the influence of a group of scholars who were more concerned with art and literature than with science. These people, called *medical humanists*, rejected changes made in the medicine of Galen and sought to return to the classical traditions of Greece and Rome. Anatomists realized that knowledge must begin with the framework: the skeleton.

Bonesetters

Bonesetters practiced for thousands of years in England, France, Germany, and Spain, long before the medical profession was organized. Sometimes they rivaled the medical profession and were held in great contempt by physicians.

Bonesetters treated not only fractures, dislocations, and sprains but also congenital disorders that were present at birth. They did not practice "in the field," like the wandering attendants that followed the troops to care for battlefield wounds. Instead, they stayed in one place and built up a reputation. They usually worked in rural areas. Sometimes their patients would come to them in large groups similar to making a religious pilgrimage.

English bonesetters were highly regarded and dealt with orthopedics before doctors as a whole took notice. They never published the secret of their art but kept it in families, handing down techniques from generation to generation. Families like the Masons, Huttons, and Bennetts practiced for generations. A Welsh family, the Taylors, practiced in Lancaster, England, for two hundred years. Some members of this family migrated to America, and their descendants practiced in Rhode Island until 1917.

THE BEAUTY OF ART

One of the most powerful forces in the advance of anatomy came from a very unlikely place—art. An Italian artist, Leon Battista Alberts (1404–1472), argued that knowledge of the skeleton conferred insight into proportion necessary for great paintings. Andreas Verrochio (1435–1488) required his pupils to study flayed bodies. Albrecht Durer (1471–1528) and Michelangelo (1475–1564) clearly expressed knowledge of anatomy in their paintings.

Most likely no one understood the anatomy of the human body better than Leonardo da Vinci (1452–1519). He would spend the night hours in the company of corpses; hoping not to be discovered, night after night he probed and cut and drew what he saw. He had no medical education, barely read Latin, and read no Greek. Dissecting over thirty men and women, he ignored previous medical writings. He was the first to picture the skeleton accurately. In his anatomical notebooks, he compared anatomy with architecture and explored the mechanics of the body. In his drawings, cords joined bones to illustrate lines of muscular tension. His artistic and scientific curiosity were unparalleled.

At age 14, Leonardo da Vinci apprenticed to Andreas Verrochio, the foremost art teacher in Florence. Verrochio insisted his pupils study anatomy. Thus Leonardo da Vinci became obsessed with dissecting for nearly fifty years. He dissected pigs, oxen, horses, monkeys, insects—and human cadavers. Contemptuous of astrology and alchemy, he also criticized physicians as being destroyers of life who lusted after wealth. It is interesting that his work had little effect on contemporary medicine because none of his anatomical notebooks were published until the late eighteenth century. His notebooks were written in a mysterious code, a form of mirror writing. Many of his projects were never completed.

With the invention of the printing press in 1450, medical drawings soon were used to illustrate texts. The first drawings were simple and childlike. Johannes de Ketham published the first medical work that included illustrations in Venice in 1464. One version showed a lecturing professor seated in a chair facing his students. He did not look at the body behind him, which was being cut open by barber-surgeons.

Geronimo Mercuriali (1530–1606) produced the first illustrated book on sports medicine in Venice in 1569. He mainly addressed health habits, sports, and exercises of the Greeks and Romans. The popular book went into six editions.

MEDICAL MEN AND ANATOMY

The literal meaning of the word *autopsy* is "to see for oneself." Indeed, the spirit of personal exploration dominated anatomy among the Renaissance physicians. Berengario da Carpi (1460–1530) lectured at Bologna, the center of dissection. His book included procedures for dissections made in public. Johannes Dryander (1500–1560) carried out public dissections at Marburg in Germany. By the 1520s, increasing numbers of anatomy texts were in print.

The pioneer of anatomy was Andreas Vesalius (1514–1564). Born in Brussels, he had a great medical heritage. His family had acquired a vast library of medical texts, and he devoured all the classics, lamenting that he needed more to read. Like other students, he accepted Galen but refused to follow him slavishly. He preferred direct observation. While at Paris as a student, he disagreed with his teacher, Jacobus Sylvius (1478–1555), and eventually ended up in Padua, Italy.

Vesalius taught himself anatomy by dissecting mice and other small animals. By 1538, he began to recognize differences between his own observations and those of Galen. Vesalius's book *The Fabric of the Human Body*, printed in Basel in 1543, is one of the treasures of medical literature. The first book presents exact and detailed descriptions of the skeleton. In this book, Vesalius corrects several of Galen's errors: For example, he points out that the human sternum has three, not seven, segments. *The Fabric* laid the foundation for observation-based anatomy, for finding facts, and for testing truth. From then on, all anatomy statements were subject to the test of human dissection.

The frontispiece of the *Fabric* depicts the dreams and agendas of the new medicine (see illustration). The cadaver is the central figure, with its abdomen

Vesalius performing a dissection. Frontispiece of *De Humani Corporis Fabrica*, 1543. © National Library of Medicine.

opened so everyone can peer in. In the sketch, a skeleton points toward the open abdomen. Vesalius looks out, extending an invitation to anatomy. Dogs and cats chase each other as the serious business of anatomy progresses. The search for truth had begun.

The judicial practice at the time was to execute criminals along the side of highways as an example to others of what could happen to them if they broke the laws. It is interesting that wherever Vesalius conducted his famous demonstrations, the bones of the executed along the roadsides and in freshly dug graves would disappear.

Although Vesalius obtained his own bodies for dissection, professors too timid to get bodies would hire enterprising entrepreneurs called *resurrectionists* or *sack-em-up-men* to do the dirty work. In fact, grave robbing became so common that George II of England declared a special Murder Act decreeing that the bodies of executed criminals be sent to the Royal College of Surgeons.

Thomas Lineacre (1460–1524), a Doctor of Medicine from the University of Padua, developed a large practice and was physician to King Henry VII of England. He and his pupil, John Caius (1510–1573), wrested medical authority away from the Church to the newly formed College of Physicians in Padua.

Just as Vesalius corrected Galen, later anatomists corrected him. Observation became paramount. Anatomists rushed to discover new body parts and then named the parts for themselves. In this way, scientists and their names became immortal.

THE HORRORS OF WAR

The horrors of war also added to the knowledge that would explode into the orthopedic revolution. Gunpowder was first used at the Battle of Crecy in 1346. In general warfare, physicians might accompany the officers but the rank-and-file soldiers had to depend for medical care on other soldiers or people who followed behind the troops. In Germany, an underclass of wanderers called *Wundartznei* roamed the battlefields patching up soldiers wounded in battle. One of these attendants, Hieronymous Braunsweig (1450–1512), believed that gunshot wounds were poisonous and needed to be cleansed by drawing silk threads through them. In his book on wound care, he said a broken bone that healed improperly must be readjusted by standing on the fracture, positioned between two wooden wedges. In 1517, Hans von Gersdorff (1455–1517) urged gentler treatment of wounds by pouring warm, not boiling, oil on an amputated stump and covering it with a flap of skin and muscle. He also used screw traction for shoulder dislocation. Around this time, many manufacturers of armor began to construct surgical appliances. Hieronymous Fabricius (1533–1619) of the University of Padua and teacher of William Harvey (1578–1657), the discoverer of blood circulation, created devices for many types of deformities from battle as well as for clubfoot, wry neck, curvature of the spine, and other conditions.

Galen and Hippocrates never faced the devastation of gunpowder. Whereas battles with swords inflicted relatively clean wounds, gunpowder maimed. John of Vigo (1460–1525), one of the first to write about the new type of warfare, believed the firearms made wounds that were poisoned just like snakebites. The treatment was to burn, or cauterize. To reach the penetrated wound, Vigo recommended pouring boiling oil on it. As wars were waged all over Europe, the field surgeons followed. Although they were unprepared in theory and schooling, they gained some knowledge through experience.

THE BARBER-SURGEONS

The barber-surgeons practiced blood-letting as well as a range of operations, including amputations. Afterwards they would wrap the bloody red rags around a pole in front of their shops to dry. This was the origin of the red and white striped pole outside barbershops. During this period, surgery was a mixture of art, science, and myth. The medical community looked down upon the barber-surgeons because they lacked formal training at universities. Instead, the barber-surgeons learned their trade as apprentices and journeymen and formed a guild, or association. From their ranks rose the most notable surgeons, who eventually led to the separation from the barbers.

English surgery of the sixteenth century was more military than civil. Thomas Gale (1507–1587) and William Clowers (1544–1604) wrote books on the subject. Clowers, who served the British fleet in 1588 when it defeated the Spanish Armada, wrote *For All That Are Burned with the Flame of Gunpowder*.

Ambrose Paré (1510–1590)

Paré was a poor French boy who grew up in the countryside around Baurg Hersent. Following his father in becoming a barber-surgeon, he worked briefly at Hôtel Dieu, a hospital in Paris, and then became an army surgeon under Henry IV. He had great compassion for his fellow men at a time when that quality was not an attribute of military surgeons.

In 1536, at the battle of Turin in northern Italy, he made a major discovery. Although Vigo had advised that wounds must be treated with boiling oil, by the end of the day Paré had run out of oil and instead made a salve of egg yolks, turpentine, and rose oil and applied it to the soldiers' wounds. He hardly slept at all that night, worrying about the men and expecting to find them dead in the morning. To his surprise, the next morning the patients were much improved and without pain. Seeing this, Paré vowed never to use boiling oil again on wounds. Moreover, he was the first to use a silk tie as a tourniquet for amputations.

In 1561, Paré himself suffered a compound fracture just above the ankle. The pain was excruciating as the attendants pulled him this way and that to get him out of danger. He instructed them to treat him just as regular soldiers were treated, encouraging the attendants to reach into the wound with their fingers to rescue bone fragments. Paré did get a major infection from this treatment, but he recovered without a limp.

In 1564, he published an encyclopedia on surgery in modern French rather than in Latin. The book gave a careful account of injuries to nerves, joints, and ligaments. Paré was the first surgeon to cut out an elbow joint to treat persistent infection; the operation was not repeated until the eighteenth century. He also designed artificial limbs, instruments, and braces. Paré endured ridicule for challenging authorities such as Galen and Hippocrates, but he practiced surgery for more than fifty years.

The dream of the Renaissance physicians was to restore medicine to its Greek purity, but unforeseen events changed their course. For example, the invention and use of gunpowder caused surgeons to look for new and better ways of treating injuries. Also, the printing press made all sorts of literature available, and people began to demand to learn to read. In these ways, the foundations were established for the "new science" of the eighteenth century.

10

The Seventeenth and Eighteenth Centuries: The Rise of Orthopedics

The seventeenth and eighteenth centuries were years of scientific revolution in the academic world, but the changes were slow to affect the daily lives of the masses. With astronomy as the vanguard, pioneers like Johannes Kepler (1561–1630), Galileo Galilei (1564–1642), and Isaac Newton (1642–1727) confirmed that the earth is not the center of the universe. Vesalius (1514–1564) had determined that the inner space of the body was a subject for exploration. Led by investigations in anatomy and physiology, surgeons and physicians began to emphasize healing and care. Meanwhile, political, religious, and economic systems were starting to undergo upheaval.

SEVENTEENTH-CENTURY MEDICINE

In the Middle Ages, dwarfs and people with extreme physical deformities had served as the basis for images of demons and devils. Later, they became objects of humor and entertainment, especially in royal courts. Gradually, however, some concerned individuals sought to understand these people's disabilities.

Ambrose Paré set the stage for this endeavor by studying scoliosis and other deformities (Chapter 17) and deciding that such conditions could be corrected by a steel corset perforated for lightness. For those who were still growing, the corset could be changed every three months. He also created boots for clubfeet and a crutch for uneven legs.

In the meantime countries in Europe, though battling each other in wars, began to share medical advances. Books translated into many languages aided in the expansion of ideas.

Great Britain

The Royal Faculty of Physicians and Surgeons of Glasgow was founded in 1599 and soon gained popularity throughout the century. In 1601, a seemingly unimportant act called the Poor Relief Act was the first European legislation to acknowledge the needs of people with disabilities and make provision for their care. Several physicians advanced not only the knowledge of bones but also the science of caring: Richard Wiseman (1622–1676), Thomas Syndenham (1624–1689), Francis Glisson (1607–1677), and Clopton Havers (1650?–1702).

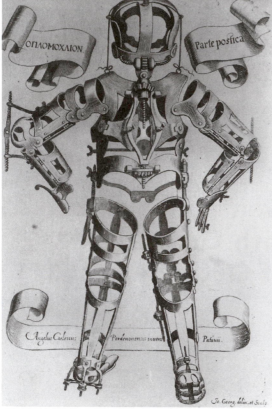

Fabricius Hieronymus Paavii, *Opera Chirurgica*, 1647. © National Library of Medicine.

RICHARD WISEMAN

When England entered a bloody civil war with rebel forces under Oliver Cromwell seeking to overthrow the Royalists loyal to the king, Richard Wiseman was in the business of amputating legs and thus was pressed into the Royalist service. Noticing white tumors that were painless but difficult to cure, he determined that swellings were related to a condition called *scrofula*, a type of tuberculosis that affects the bones. The condition was also known as the King's Evil because it was supposedly cured by a touch of the royal hand. Hippocrates (460–370 BCE) had earlier proposed that tuberculosis (a disease of the lungs) and a certain condition of the spine (also called *scrofula*) were in fact manifestations of the same disease. Wiseman studied the two diseases and convinced many of his fellow doctors that the two conditions had the same cause.

THOMAS SYNDENHAM

Born in a small English village, Thomas Syndenham emerged as one of the barber-surgeons who was influenced by Hugh Owen Thomas, a bonesetter who had lived two centuries earlier. Syndenham had a great intolerance for blind authority, but he trusted the natural processes pronounced by Hippocrates and emphasized clinical observation. As a sufferer of gout (see Chapter 15), he described this condition as well as rheumatoid disorders. His questioning and dissent caused him to be given the title "The Father of English Medicine."

FRANCIS GLISSON

In the early years of Glisson's life, no one was interested in treating deformities. Imposters and charlatans did not want to treat them because there was no quick treatment for profit. But as an adult, Glisson became interested in rickets (see Chapter 15). In the past people with the condition had been fitted with padded iron corsets, but Glisson introduced orthopedic exercises, massage to counter weaknesses, and a suspension device that could lift the body weight off of the legs. His book described rickets in such great detail that it became known as the English disease.

CLOPTON HAVERS

As a young man this Englishman went to Utrecht, the Netherlands, and came in contact with Anton van Leeuwenhoek (1632–1723), who had described microscopic vascular channels entering the bones and the crosswise channels within the bones (see Chapter 5). Havers subsequently used the microscope to look at many of the structures of bones and joints. In 1689, he gave a series of lectures to the Royal Society dealing with the fine architecture of the bones. The Haversian systems bear his name. Havers laid the groundwork for the study of physiology and pathology of the bones.

EIGHTEENTH-CENTURY MEDICINE

The years 1700 to 1799 are important to medical history because during that time surgeons developed better techniques for amputations as well as other important and complicated surgeries. Without the benefit of anesthesia, speed was important. Amputations were performed while strong-armed assistants held the patients down. The great amputators controlled bleeding and infection with whatever materials were available and developed some unique antiseptics.

Amputation

Topping the list of achievements for amputation was the tourniquet. The French surgeon Jean Louis Petit (1674–1750) developed a tourniquet that involved twisting down a screw compressor device on the patient's leg and on the lower abdomen. This stopped the blood flow in the main artery and enabled surgeons to perform amputations higher up on the leg than previously. Many consider this the most important surgical advance before anesthesia. Speed became a source of pride. The French surgeon Jacques Lisfranc

(1790–1847) could amputate at the thigh in 10 seconds. The hospital mortality, or death, rate was 45 to 65 percent.

Fractures

After falling from his horse, English physician Percival Pott (1714–1788) had a painful ankle injury in which his foot turned outward and backward.

Andry de Bois-Regard, Nicolas, *L'orthopedie* . . . , Paris, 1741. Andry declared that the same method must be used to shape the leg just as the young tree is reformed by straight poles. © National Library of Medicine.

He refused to be moved until someone purchased a door for him to be carried on. He was convinced that the jostling and jolting of a carriage ride would make the injury worse. With such an injury, amputation was usually performed immediately; but Pott said no. And, indeed, the fracture healed itself. This fracture—at the end of the fibula and middle part of the lower tibia—now bears his name: *Pott's fracture.*

Pott also investigated paraplegia, or paralysis of the lower portion of the body caused by spinal tuberculosis. He was convinced that it was not a condition of the spinal cord but was caused by the lung disease tuberculosis. Because of his lifelong interest in and study of the disease, the paralytic disorder was named for him: *Pott's disease.*

The Word Orthopedics

In 1741, at an elderly age, Nicholas Andry (1658–1747), professor of medicine at the University of Paris, published *Orthopaedia: The Art of Correcting and Preventing Disabilities in Children* and included a now-famous illustration (see illustration). In explaining his title, he stated that he had formed it from two Greek words: *orthos,* which signifies "straight"; and *paidion,* meaning "child." Out of those two words he compounded *orthopaedics* to teach methods of preventing and correcting disabilities in children. Under the picture he wrote, "The same method must be used . . . for recovering the Shape of the Leg, as is used for making straight the crooked Trunk of a young Tree."

Some people say he is the father of orthopedics, but others say he only gave it a name. The rest of his book is a medical "gossip column" on complexion treatments and behavior.

Thanks to Andry, a generation of schoolchildren was admonished to sit up straight in hard-bottomed, straight-backed chairs. He believed that perfect posture in straight chairs was the only way to prevent back problems in adults. He developed braces, splints, and other devices for a number of orthopedic problems.

The Barber-Surgeons

The barber-surgeons continued practicing during the eighteenth century, but some became convinced that surgery should be a separate profession. Several of the following surgeons worked to make surgery a true profession.

WILLIAM CHESELDEN

At age 15, William Cheselden (1688–1752) was apprenticed to the barber-surgeon William Cowpens (1688–1709) for seven years. After his talent for surgery was recognized, he taught the first regular course for surgery in 1710. At the time, surgeons did not study formally but worked with an older barber-surgeon as an apprentice. In 1713, Cheselden wrote *Anatomy of the Human Body*, which was used as a text for over one hundred years. A later book, *Osteography,* had fifty-six fine copper plates that showed life-size bones. Cheselden worked hard to get surgeons recognized as separate from barbers. (See "The Surgeons and Barbers Part Company.")

Cheselden was a caring and humane physician. In one of his books, he

The Surgeons and Barbers Part Company

The guild, or union, system for barbers and surgeons developed in the Middle Ages. In 1744, surgeons held a meeting in London expressing a desire to separate from the barbers. But the barbers organized a campaign opposing the separation. When the proposal was submitted to Parliament, physician Charles Coats headed the committee hearing the proposal. It so happened that Coats's father-in-law was the surgeon William Cheselden, who greatly influenced the decision. So, in May 1745, Parliament passed the bill separating the two. The Company of Surgeons held its first meeting in 1745. However, the medical community still did not respect them until the nineteenth century.

told how he had difficulty sleeping the night before a difficult surgery. In another well-known story, a bonesetter treated Cheselden's broken elbow with linen strips soaked in egg white and flour. Cheselden then began to use this method himself—on both fractures and congenital deformities. Children with clubfeet were brought to him from many places, both near and distant.

He was the first to do surgery for torticolis, a deformity of the neck wherein the shortening of the neck muscles causes the head to tilt to one side. He performed a tenotomy—a cutting of the tendon—to relieve the condition.

ROBERT CHESSER

When Robert Chesser (1750–1831) was a boy, his mother married a surgeon who developed a successful practice treating fractures and accidents of coal miners. Sometimes the stepfather left the treatment of fractures to the boy. After an apprenticeship with an obstetrician, Chesser devoted himself almost exclusively to fractures and deformities. His very large practice mostly included children. He had his own workshop with six workmen, who created splints and other appliances. In 1617, he helped form the General Institute for the Relief of Persons Labouring under Bodily Deformity. It was renamed the Royal Orthopedic and Spinal Hospital in 1925. Chesser was the first fulltime orthopedist.

Inquiry into Disease

JOHN HUNTER

John Hunter (1728–1793), a British physiologist and surgeon, was born in Scotland. His inquiring mind and enthusiasm for learning had a profound effect on gaining respectability for surgeons. A farm boy, he did not begin his apprenticeship until age 20. He then worked in his brother's dissecting room and became a surgeon during the Seven Years' War with England and Prussia fighting a coalition of countries led by Austria and France. As a researcher and lecturer, he had many esteemed pupils such as William Chesser and Edward Jenner (1749–1823), the developer of the smallpox vaccine. Although Hunter had little schooling, he was one of the first to put surgery on a scientific foundation and lay the groundwork for nineteenth-century development. His saying, "Don't think—try the experiment," inspired generations of surgeons.

Hidden away at Earl's Court, a rural area of London, he experimented on animals. He determined how to assess muscle power in a weak muscle. Regarding joint injury and disease, he stated that voluntary movement should not be permitted until inflammation is relieved. He believed the surgeon's job was to assist the body's innate power. Using animal studies, he traced the development of new bone and studied bone absorption.

Hunter has often been compared to Leonardo da Vinci with his inquiring mind and adventurous spirit as well as his enormous influence on modern biological concepts and pathological processes that affect bone. He described the plastic nature of bone. Using animal experiments with dye and other materials, he demonstrated the continuous remodeling of layers as bone grows. He also made attempts at tissue grafting.

He collected over 14,000 specimens of animal, plant, and human tissue, from the simplest to the most complex. His collections were partially destroyed during the bombings of World War II and are now housed in the College of Surgeons, London.

JEAN-ANDRE VENEL

Because the care of children with disabilities had never been important in the past, there were no homes or hospitals to treat them. Jean-Andre Venel (1740–1791) established the first hospital for children. Venel, a Swiss physician who studied anatomy at Montpelier, determined that children needed a true hospital where they could be managed by medical staff. Braces and appliances were made there for individual children; education for children and vocational training for adolescents were also available. The hospital, or institute, treated conditions such as clubfoot, tuberculosis, and scoliosis. Breaking with some of the attitudes of the past, Venel recorded and published his methods. He also was aware of the benefits of sunlight and its relationship to health.

Venel might be called the true father of orthopedics. His institute established a model for the many institutes that developed throughout Europe in the 1800s.

OTHER PIONEERS

To ease disabilities and prevent problems, Pieter Camper (1722–1789), a Dutch anthropologist and anatomist, became concerned about the nature of shoes and their effect on the feet. People did not realize the importance of wearing shoes that fit. At the time, the Dutch were famous for their wooden shoes, and growing children would inherit older siblings' shoes when they outgrew them. In 1781, Camper wrote *Dissertation on the Best Form of Shoes*, in which he emphasized problems arising from the relationship between the mechanics of the feet and the nature of shoes.

It sometimes takes a personal accident to spur progress in a particular field. In 1773, William Hey (1736–1819) banged his knee while getting out of the bath. Thereafter the knee became his obsession and chief concern. Born in Leeds, England, he had apprenticed at age 14 to a surgeon and apothecary. He noted that bonesetters could cure knee problems by manipulation, and he studied what happened to the lesions when they were dissected. His dissections revealed that traces of meniscus, or cartilage, in the knee moved around or were displaced when an injury occurred. Hey worked out a method of putting the ligament back in place by gradually extending the knee and then suddenly flexing the joint until the cartilage moved back into place. He was the founder of surgery at Leeds.

A hallmark of the eighteenth century was an interest in the growth and structure of bone. Studies by several scientists who added dye to animal feed were a starting point. At Guy's Hospital in London, John Belchier (1705–1762) studied bones and noted that red dye was absorbed. Albrecht von Haller (1708–1777), a Swiss physiologist, and Henri-Louis Duhamel (1700–1782), a French surgeon, came to different conclusions. Haller thought the bloodstream carried bone to the periosteum, where it grew. Haller was known as the father of experimental medicine. Duhamel argued the bone grows from the periosteum very much like a tree grows from its bark. These theories sparked rival schools of thought that conflicted well into the twentieth century.

The seventeenth and eighteenth centuries produced great teachers of surgery who demonstrated their art and science not only to train medical students but also to educate the general public. Surgical amphitheaters came into their own in the seventeenth century, making way for hospitals in the eighteenth century.

Eighteenth-century surgeons were aided by hospital reformers who demanded clean beds and fresh air. One of the most notorious hospitals was Hôtel Dieu in Paris, where it was not uncommon to find five or six people in a bed—several of whom were dead. The rise of the specialty institute of orthopedics was a result of new concern about people with disabilities.

These two centuries saw remarkable change. The surgeons and researchers had students who practiced well into the nineteenth century, when other beneficial discoveries occurred. Anesthesia and humane hospital practice carried bone science and orthopedics to the edge of modern application.

The Nineteenth Century: Orthopedics as a Specialty

Whereas the eighteenth century was known as the childhood of medicine, the nineteenth century could be called its adolescence. Although advances did take place in the previous century, it was only in the nineteenth century that medicine became truly scientific. The integration of basic science and systematic research with clinical observation and autopsy promoted forward movement. Investigations into the nature of bones improved orthopedic surgery. Effective measures for controlling pain and preventing infection were key elements in the transformation of orthopedics.

This is the century when surgery broke through old barriers, including the specialty of orthopedics. More change occurred in the nineteenth century than in the previous two thousand years of recorded history. In 1800, surgery rarely went beyond fractures and amputations. At the end of the century, well-trained practitioners were graduating from institutions in Paris, London, Edinburgh, Vienna, and Leiden. The merging of surgery and medical education enhanced basic science. The many battlefields of Europe gave young orthopedists boundless practice.

THE BRITISH ISLES

Several Scottish physicians specialized in the skeletal system. John Reed (1808–1848), fascinated by the knee, was the first to describe the nature of a torn meniscus and how fibrocartilage developed on its edges. John Goodsier (1814–1867) identified certain cells, which he called *corpuscles*, in the soft hyaline membrane of the bones. But another surgeon, Sagenbauer, actually called these cells *osteoblasts*, and that name stuck. A famous German researcher, Rudolph Virchow (1821–1902), had acknowledged the important contributions of cellular function to tissue, but Goodsier linked the soft osteoid, or "bone-like," part with cartilage and connective tissue. Peter Redfern (1821–1912), a student of Goodsier, showed in animal experiments that tears in articular cartilage were healed by fibrous tissue but that the new tissue was not the same as the original. James Syme (1799–1870), a professor of surgery at Edinburgh, performed a type of quick amputation that was given his name. It was commonly performed during the American Civil War and in World Wars I and II.

One of the greatest Scottish surgeons was James Paget (1814–1894). He was a practical man who encouraged his students to imitate the best of the bonesetters and avoid the worst. He was fascinated that bone could repair itself and studied tumors of bone. In 1877, he first described *osteitis deformans* (see Chapter 17), a condition involving increasing bone tumors and head size along with other deformities. This condition is now known as *Paget's disease*. (It is thought that this disease caused deafness in the composer Ludwig van Beethoven.) Paget was a marvelous lecturer and one of the first specialists in bone pathology.

Joseph Lister (1827–1912) advanced the future of surgery by helping control the problem of infection. He discovered that antiseptic techniques could promote healing by controlling life-threatening infection. He applied carbolic acid in the form of creosote to a compound fracture on August 12, 1865—the first use of an antiseptic treatment. Writing in the publication *Lancet*, he reported that when he arrived at Edinburgh Infirmary in 1854, the amputation mortality rate was 43 percent; with his treatment, it fell to almost zero. German and French physicians were enthusiastic about his findings. In fact, Richard von Volkman (1830–1889) from Halle, Saxony, was the first in Germany to adopt Lister's methods.

Hugh Owen Thomas (1834–1891) is considered the grandfather of orthopedic surgery. Along with his pupil, Robert Jones (1858–1933), he added to current knowledge about the skeletal system and applied it to surgery. Thomas came from a family of Welsh bonesetters that went back over four generations. As a child he went with his father to Liverpool, England, where the established medical profession hated the bonesetters. The bitter attacks made the young man very defensive, but ultimately he prospered and patients came from near and far to see him—even from America.

Thomas believed in the value of rest for treating bone disease, which at the time was mostly caused by tuberculosis. He abhorred amputation and thought it unnecessary in most cases, believing with Thomas Syndenham (see Chapter 10) in the importance of natural healing. In 1859, he separated from his father after a violent argument and purchased a house that had room for eight patients. His workshop and examination rooms were in his own home. Thomas began his rounds at five o'clock in the morning, saw patients until two in the afternoon, and did surgery in the later afternoon and evening. He often worked on splints or wrote books well into the night.

His peculiar habits set his solitary life style. Always wearing a sailor's hat cocked over one bad eye, (see photo in color section) he never left home except to doctor someone. He never went on vacations or used his academic degrees to develop a career as a lecturer, although he did attend lectures by top surgeons in England and France. He never joined the staff of a hospital, and his books were published locally and not widely read.

Because Liverpool is a port, Thomas's practice included sailors from all parts of the world who had all sorts of old and problem fractures. He devised methods of treating poorly healed wounds, had lots of success, and let everyone know it. An average week included twenty major fractures, bone diseases, and many cases of joint disease and deformities. He treated as many as eighty patients a day in his home and then visited the homes of those who could not come to him. Sunday was a free day in which people came from many areas for treatment. He was committed only to his patients and did not fraternize with people or other doctors. A friend of the Thomas family, Benjamin Brodie (1786–1862), published Thomas's popular reference book in 1819. Thomas described chronic bone abscesses that later were named for him. In 1813, he made a lasting contribution to medical excellence by encouraging the Fellowship Examination of the Royal College of Surgeons to improve the education of surgeons and gain respectability for his branch of medicine.

Robert Jones (1857–1933) was a pupil and colleague of Thomas who was completely committed to his principles. A kind and well-organized man, he fostered cooperation among orthopedic surgery, charitable care, and government action. In 1895, Jones heard of a new box that used a kind of ray that could look through the skin and "see" bone. He immediately recognized the value of such a device and ordered one for his practice. The device was the product of Wilhelm Conrad Roentgen (1845–1923). (See "Peering into Bones: Roentgen's New Ray.")

In 1904, Jones confined his practice strictly to orthopedic surgery; later, he became better known in the United States and parts of Europe other than in England. However, throughout the next century both Jones and Thomas exerted enormous influences through their students.

GERMANY AND THE CONTINENT

German scientists made enormous contributions to orthopedic science and practice. Much originated in and around the town of Wurtzberg, in addition to Roentgen's X-ray. Johann Georg Heine (1770–1838), a manufacturer

Peering into Bones: Roentgen's New Ray

A physicist in Wurtzberg, Germany, Wilhelm Conrad Roentgen (1845–1923) was working one day with a vacuum tube developed by Sir William Crookes (1832–1919). The charges of electricity made the tube gleam and flicker, casting a greenish light across the room. Roentgen darkened the room and wrapped the Crookes tube in black cardboard to screen out the bright light that it emitted. Then he became interested in the light, or glow, which was produced when high voltage is applied between two wires sealed in a vacuum. Noting that a small object on his workbench glowed after being accidentally in the greenish light, he found this object was actually a small fluorescent cardboard screen that he had coated with barium platinocyanide for an unrelated project. In addition to cathode rays, the tube was evidently producing other rays that were passing through the cardboard covering the tube and hitting the coated screen.

Experimenting with the tube, he observed that the agent he called *x-rays* (because they were unknown) could penetrate books, wood, and glass and produce a picture on treated paper like the object on his bench. The only thing the rays could not pass through was lead. On December 22, 1895, he took a picture of his wife's hand, thinking it would be a unique Christmas present.

When Roentgen published his results in the scientific journal *Nature*, there was an immediate furor. Realizing the implication for medicine, energetic doctors like Robert Jones went to Wurtzberg to purchase x-ray machines. Researchers began to see many possibilities. Some experimented with picturing the esophagus. Another doctor, Georg Perthes (1869–1927), used x-rays to treat **malignant tumors** and began the field of radiation therapy.

But there was a down side to the discovery. Some outlandish claims were made that x-rays might be able to look through women's clothing. One store in London even advertised x-ray-proof underwear. A much more serious down side was that the rays might cause cancer, although the positive connection was not made until the early part of the twentieth century. Nonetheless, for this very important discovery Roentgen received the Nobel Prize in 1901.

of instruments, founded the first institute for orthopedic practice in Germany. The institute survived until World War II. His nephew, Bernhard Heine (1800–1846), invented the *osteotome*, a chisel beveled on both sides for cutting through bones; he also conducted important studies on bone regeneration.

In Berlin, Julius Wolff (1836–1902) became fascinated with the bone studies of Belchier, Hunter, and Duhamel (see Chapter 5). With no x-rays, Wolff developed a technique for cutting a thin bone section to study the trabecular structure. He found that bone deformities led to changes in the internal structure of the bone. He saw that if normal bone is used in a new way, an adaptive change takes place in form and pattern; he also saw that a deformed bone that is corrected regains its normal shape. Wolff's law states, "Every change in the form and function of a bone is followed by specific definite changes in its internal architecture in accordance with mathematical laws." Wolff's law is actually the law of osteoblasts. Wolff was also a visionary in other ways. For example, he predicted the transplantation of allogenic bone, or bones of other humans, and pioneered many ideas of surgical techniques. He wrote a famous book, *The Law of Bone Transformation*, that related form to function.

When Guillaume Dupuytren (1777–1835) was a young boy, a rich woman kidnapped him because he was so good looking and took him to Toulouse, France. After he was rescued he was sent to Paris to study medicine, where he endured great poverty and allegedly used the fat from cadavers to fuel his lamp. In 1813, he became chief surgeon at the hospital Hôtel Dieu in Paris. Although he was rude and harsh to patients, he reigned as absolute monarch of surgeons in France for twenty years. An accurate clinical observer with an interest in pathology, he was the first to classify burns systematically and also developed plastic surgery for repairing skin defects and scars. He characterized and treated a contracture of the hand that was named after him, *Dupuytren's contracture* (see Chapter 16).

A Dutch military surgeon, Antoine Mathysen (1805–1878), invented a kind of plaster of Paris (POP) bandage in 1851. A debate of the pros and cons of using it ensued. Many doctors who believed that nature should have a part in healing opposed the bandage as being too cumbersome. But Mathysen considered this bandage, which had been used in a crude form in earlier centuries, ideal because it set rapidly, gave easy access to the injured part, and was not too heavy or extensive. POP ultimately gained universal popularity and helped people who would have been bedridden to move around.

THE UNITED STATES

Compared to Europe, the United States provided a completely different setting for the development of orthopedic surgery. Two factors affected its development there:

- The ocean isolated the United States from immediate influences of entrenched medical institutions elsewhere.
- Many immigrants influenced and encouraged a fresh new approach.

Although many ideas did come over from Europe, Americans developed new ones with energy and thoroughness. Indeed, the development of orthopedic surgery in the United States as an independent discipline was faster than in Europe.

At the time of the American Revolution, most medical men worked in the northern British colonies. The only two medical schools were in Philadelphia (established in 1765) and New York (established in 1767). Before the Revolution, these schools had awarded fewer than fifty medical degrees. In the previous wars that had been fought in America, most campaigns had involved Native Americans, who had limited access to guns and fought only on the frontiers. Thus the colonies had not seen massive casualties like the ones the wars in Europe produced. But the British in the Revolutionary War brought guns that used gunpowder and inflicted serious wounds that became infected. U.S. doctors had a new challenge to treat these injuries.

John Mann, a U.S. physician, studied gunshot wounds. In 1816, he wrote *Medical Sketches of the Campaigns in 1812*, an early manual on amputation. Because medical education in the United States was limited, some colonists went to England and the Continent to study medicine.

One of the early American surgeons was Valentine Mott (1785–1865). He studied abroad and brought back some refined techniques; he was also the first to use the term *orthopedic surgery*. Boston provided a proper climate for his new specialty, and he persuaded a colleague, John Ball Brown (1784–1862), to join him. Thus American orthopedics as a specialty started in Boston. Brown's son, Brickmeister Brown (1819–1891), published the first American textbook and was a founder-member of the American Orthopedic Association in 1887.

The use of anesthesia aided in the development of surgery. Henry Jacob Bigelow (1818–1890) was the first to employ ether at the Massachusetts General Hospital on October 16, 1848. This was heralded as "ether day." Bigelow had heard how T. G. Morton (1819–1868) had used ether to extract teeth. So Bigelow persuaded the famous surgeon John C. Warren (1778–1856) to allow Morton to give one of his patients ether during surgery. A few weeks later, Bigelow described the possibilities of anesthesia in the role of surgery in a published paper. This occurred only a few years after a surgeon in Europe asserted that to eliminate pain during surgery was not practical and should not be pursued.

Ether was soon replaced by chloroform, which was discovered in 1831. Tradition holds that James Young Simpson (1811–1870; see photo in color section), a professor of surgery at Edinburgh, was testing chemicals when someone upset a bottle of chloroform. When his wife came in to bring his supper, she found the laboratory staff asleep on the floor. A key event happened on April 7, 1853, when Queen Victoria took chloroform during the birth of a son.

The "Father of American Surgery" was Philip Syng Physick (1768–1837). The son of an English immigrant, he became a pupil of John Hunter in

Leicester Fields and was his close assistant in animal experiments. As a professor at the University of Pennsylvania in 1816, he found that the body could absorb ligatures made from animal tissues. Both he and his assistant, John Rhea Barton (1794–1871), were innovative and aggressive. Barton was equally skilled with either hand; once he assumed a position for an operation, he never moved. Barton is best known for innovative osteotomies, or cutting out of bone, and treating a fracture of the wrist now known as *Barton's fracture*.

In the United States, sports medicine started with Edward Hitchcock (1793–1868). In 1854, he became a physical education teacher and hygiene instructor at Amherst College in Massachusetts. He is considered the founder of the American physical education movement, the first sports medicine specialist, and the first team physician.

The American Civil War: Medical Dark Spot

The one dark spot in American medicine during the nineteenth century was the Civil War (1861–1865). Neither side was prepared for the conflict, which both sides thought would be over in a matter of months. Both sides predicted victory. The national army was not adequate to meet military needs. West of the Mississippi, there were great demands relating to critical situations with Native Americans that had developed because of the Gold Rush to California. Thus medical organization and equipment were already in a crisis before the war among the states even began.

At the beginning of the hostilities, the Medical Department had one surgeon-general and a force of 115 trained medical officers (see photo). Many of the members were Southerners, and 27 of them resigned to become the Medical Department of the Confederacy. They began to recruit from civilian doctors to meet the demands in vast, spread-out battlefields. So great was the need that more than 12,500 physicians

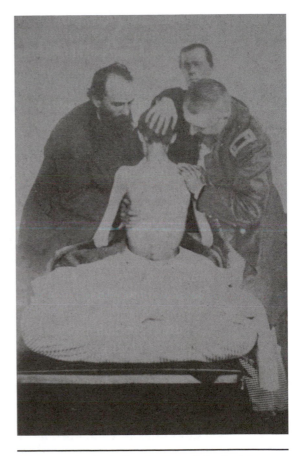

Doctors in the Civil War examine a federal prisoner returned from prison, 1861–65. Courtesy of the Library of Congress.

from the North and 3,000 from the South were recruited. There were also a host of unknown civilian medical volunteers who were not distinguished from trained doctors. After every engagement, these volunteers would flock to the site to "help out." Many of them were imposters, cultists, and people who lacked military discipline.

At the beginning of the war, each regimental surgeon was given equipment in a cumbersome surgical knapsack that weighed about 20 pounds. Surgeons expected infection. They also expected recovery to be a slow, painful, and exhausting process. No antiseptics were provided. Whereas Napoleon's armies—as well as those of many other countries in Europe—had developed dressings, mobile ambulances, and sophisticated treatments, American Civil War medical procedures involved the cleaning of wounds with dirty handkerchiefs or pieces of cloth torn from sweaty shirts. Medicines in pill form were awkward to administer. Although surgical anesthesia was a new procedure, most Civil War–related operations were performed without anesthesia.

Medical officers on both sides were charged with incompetence, ignorance, inefficiency, neglect, cruelty, and drunkenness. They often performed unnecessary operations or made wounds worse as they dug around for embedded bullets. More soldiers died during this conflict of cholera, dysentery, and gangrene than from actual battle wounds.

Surgical Wards Incorporate the Old and the New

Late nineteenth-century surgical wards displayed a combination of the old and the new. The surgeon would operate in street clothes and did not scrub his hands before surgery. He would throw his coat on a chair, roll up his sleeves, and put on a blood-stiffened coat. Even Lister, the discoverer of antiseptics, only rinsed his hands with carbolic acid before surgery. In 1897, the Polish surgeon Johannes von Mikuleiz-Radiki (1850–1905) realized a surgeon's breathing out of germs might contaminate the patient; he was the first to use a mask. Berkeley Moynihan (1865–1936) developed an additional aseptic technique: He was the first to wear gloves.

The heyday of American-made instruments, from the 1840s to the 1890s, was centered in New York and Philadelphia. Before this, most instruments were made in Europe. Many were made of ivory, wood, or other materials. However, these materials could not withstand the steam of sterilization. After 1890, only metal instruments were used.

As orthopedics emerged as a specialty, special attention was paid to spinal deformities in young people. The development of hospitals and the rise of the nursing profession were also part of nineteenth-century medical progress. At the beginning of the nineteenth century hospitals were only places for care of the poor. Unsanitary conditions and poor facilities made them undesirable. But by the 1880s as surgeons began to accept aseptic conditions. Coupled with the use of nursing as a profession, hospitals turned from refugee for the poor into institution fit for everyone.

Medicine in the nineteenth century continued in the hope of the Enlightenment of the eighteenth century. Although there were setbacks like the dark blot of the Civil War, by the end of the century, advances in the medical profession were setting the stage for the technology and advances of the twentieth century.

The Twentieth Century: Bones, Research, and Advancing Technology

Tendon transplantation, bone grafting, and other procedures marked the early 1900s as a turning point for orthopedics. Indeed, many improved surgical practices had been discovered by the turn of the century, but they were not necessarily widely accepted. The idea that germs, or microbes, caused diseases had to be proved. The precautions that were required during surgery had to be demonstrated. Dedicated doctors and surgeons helped to promote the discoveries and put them into practice.

The beginning of the twentieth century found many ideas from earlier practitioners take hold and develop:

- Joseph Lister (1827–1912) had promoted asepsis using carbolic acid, or phenol, to sterilize wounds and instruments.

- The theory that germs caused disease began to gain acceptance.

- Surgeons began to don hospital gowns instead of wearing street clothes.

- The German physician Ernst von Bergman (1836–1907) introduced the steam sterilizer for tools used in surgery.

- The German surgeon Herman Kummel began the practice of carefully washing the hands before surgery, while William Stewart Halstead (1852–1922) ordered special rubber gloves. In 1896, the gauze mask was introduced.

- The anesthesias chloroform and ether used at the turn of the century were popular, but by 1920 nitrous oxide, or laughing gas, was combined with ether. In 1920, Ralph Waters (1883–1979) of Madison Wisconsin, introduced cyclopropane. In 1933, John Lundy (1865–1959) of the Mayo Clinic in Rochester, Minnesota, used pentothal. The poison curare was used for local blocks.

- Shock, a longtime problem for surgery known to be caused by loss of blood, was addressed by Karl Landsteiner (1868–1933) when he discovered the blood factors A, B, AB, and O. Transfusions would now be possible by matching blood types and without causing deaths by mixing incompatible types.

MEDICAL ADVANCES DURING WARTIME

A significant medical benefit from the two world wars during the first part of the twentieth century was the advancement of orthopedic surgery. Valuable medical lessons that advanced the understanding of bone structure and treatment were gained on the battlefield and later applied to civilian life. Before firearms were invented, combat with sword and spears produced little damage to bone. But the high-powered projectiles of twentieth-century warfare permeated tissue, and groups of bacteria called *anaerobes* began to cause serious problems. *Anaerobe* means "without air." Sealed in without air, these bacteria multiplied in the body and released deadly toxins that caused fatal tetanus and gangrene, diseases caused by deep penetrating wounds. In previous wars, the only answer to such problems had been amputation of the infected limb.

American Red Cross personnel holding a wounded soldier on a stretcher at the Italian front during World War I. Courtesy of the Library of Congress.

World War I (1914–1918) broke many medical traditions. Foremost was the abandonment of surgical treatment in hospitals for impromptu treatment on the battlefield. Young doctors who had been trained in clean, antiseptic hospitals were pressed into action in grimy field facilities (see photo).

In 1915, Edward Milligan, an Australian serving in France, encouraged debriding (cleaning dead skin) and simple incision or draining as a way to prevent gas gangrene, one of the two types of gangrene. Alexis Carrel (1873–1948) and Henry Dakin (1880–1953) developed a treatment in which sodium hypochlorite (bleach solution) was used after debridement. These treatments reduced the number of amputations. The war indirectly accelerated progress as surgeons began to pool their knowledge, publishing their research in developing journals and putting what they read there into practice. Fortunately, the new knowledge carried over into civilian life.

As the x-ray slowly gained in popularity, doctors could look below the surface to see conditions such as osteochondritis and osternecrosis (see Chapter 17). In 1898, George Clemens Perthes (1869–1927) took the first x-rays of what later became known as Perthes disease, although he did not publish his findings until 1914.

World War II provided another opportunity for doctors to expand their knowledge. Building on lessons learned in World War I, army physicians performed fewer amputations, saw less gangrene, and devised ways to fix, or set, fractures. A highlight of this period was the discovery of penicillin by Sir Alexander Fleming (1881–1955) in 1928 (see "Penicillin: A Great Discovery of the Twentieth Century"). When it was first used in North Africa, the incidence of infection—which had been a major cause of disability—was reduced tremendously.

Seeking to send injured soldiers back to the front as soon as possible, the Germans developed a number of procedures that involved inserting special nails into wounds to hasten bone healing by holding the fracture in place. Gerhard Kuntscher (1900–1972), a German doctor, published details about this revolutionary procedure in the opening months of the war. His work involved the intramedullary nailing of fractures of the shafts of long bones. Today, this special nail carries his name.

IMPLANTS

Many surgeons used metals such as brass, along with catgut from camels, to replace bone and tendons, respectively, but usually these techniques were not successful. In 1877, Joseph Lister (1827–1912) wired the patella without success. However, his methods of asepsis gave the impetus for increased efforts to try internal metal fixation. Initially, implants with silver, steel, and platinum, along with silk, met with the same disappointment.

Search for Metals

The hunt to find the best materials for implants was on. Iron and carbon steels dissolved, affecting the bone. Inserts of copper, nickel, zinc, and aluminum alloy discolored the bone. There was no reaction to gold or silver but they did not prove feasible. Lead proved inert and corrosion-resistant but was later shown to be poisonous. Then, other materials came to the forefront. For example, carbon fiber reinforced with epoxy resin could be used as plates fixed

Penicillin: A Great Discovery of the Twentieth Century

In late summer of 1928 in a crowded, dark laboratory at St. Mary's Hospital, London, Alexander Fleming (1881–1955; see illustration in color section) was working on a bacteriology textbook that required him to prepare cultures of the bacteria *Staphylococci*, sometimes called *staph*. This type of bacteria causes boils and other pus-producing infections. Under the microscope, they appear round like a bunch of grapes.

Fleming was on summer vacation but had to come into the lab to examine a number of culture plates and dispose of them. One plate caught his attention. A plate dotted with colonies of staph had a large mold growing in it. In the area around the mold, the staph were transparent and almost ghost-like; in fact, the bacteria around the mold were dying. When Fleming showed the plate to his colleagues, they were not interested because many of his previous experiments had been failures.

But Fleming was not discouraged by his colleagues' reactions. As the mold continued to fascinate him, he took a picture of the plate and preserved both the mold and the staph with a chemical. He ultimately found that the mold was a member of the *Penicillium* family, so he called the mold juice *penicillin* after its family name. However, because the bacteria-killing property could not be put into an extract, there was little reason to continue research on it.

Howard W. Florey (1898–1968), an Australian-born scientist, became interested in Fleming's work during the 1930s. He recruited an enthusiastic and skilled biochemist, Ernst Chain (1906–1979), who put aside his original investigation of snake venom to research the killing power of Fleming's mold. Because Great Britain was at war with Germany at this time, there was no grant money to fund research on mold. So Florey and Chain came to the United States, and convinced a drug manufacturer there of the value of producing penicillin. It was first used to treat mass victims of a terrible fire in Boston on November 28, 1942. From there it was sent to the battlefields of World War II.

with conventional screws of metal. The carbon fibers could also substitute for tendons and ligaments, particularly of the knee. Biodegradable fixation systems that allowed tissue growth were among other experimental methods.

After World War II, Robert Danis (1880–1962) of Brussels was dissatisfied with the current fixation techniques and introduced the idea of compression. Subsequently, three Swiss workers, noting the processes in a watch-spring factory, applied Danis's ideas and developed lag screws, compression plates, and tension band wires inserted with special tools. Studying the effect on bone healing, they preferred this method to unyielding devices such as rigid plates.

Bone for Bone

In 1682, Job van Meekren in Holland had filled a defect in a soldier's cranium with a piece of dog skull, but the Church had considered it improper and made him remove it. Three centuries later, substituting one bone for another became a popular idea. Fred Albee (1876–?) grafted bone in New York in 1911. He fitted a cortical graft as an inlay: periosteum to periosteum, medulla to medulla, and cortex to cortex. Albee also found that chilling bone preserved it for later surgery.

Another type of bone substitution involves an *autograft*, in which the person gets his or her own bone. A major drawback is that the receiving bone might recognize the grafted bone as its own and resorb (absorb again) it, leaving a hole. If this happens, it might require a long series of grafts before one takes. Also, there might be scarring and infection in the area where the donor bone was harvested. An *allograft* is a transplant from a member of the same species, which avoids resorption of the recipient bone. Cadaver bone can be freeze-dried and stored at room temperature with vacuum packing for this purpose.

Bone marrow transplants involve taking marrow (usually from the hip) and using it to promote new growth. E. Donnall Thomas (1920), of the University of Texas won the Nobel Prize in Medicine in 1990 for research in this area. In his Nobel lecture he explained how marrow grafting has progressed from a highly experimental procedure to clinical use in a variety of diseases. He also projected the future of bone marrow transplants with stem cell research (see Chapter 13). In addition, he emphasized that marrow grafting never would have been possible without animal research using mice and dogs.

Biomechanics: The Application of Principles of Mechanics to the Living Organism

Three important factors came together to make implants that work: the development of metals and materials, the availability of potent antibiotics, and the understanding of the mechanics of the musculoskeletal system. Early experiments like those of Theophilus Gluck had tried to address the problem without understanding how bones work and how to control infection.

In the nineteenth century, rubber, ivory, inorganic materials, and synthetics were used for implants. Subsequent advances in chemistry introduced plastic polymers such as polyethylene, polypropylene, epoxy resins, and polyurethane. Silicones are chemical molecules that attach a silicon atom to a carbon chain to make a polymer. Polymers are usually used in finger joint implants and the replacement of carpal bones.

In 1960, George McKee and John Watson-Farrar (1926–1999), English surgeons, attempted a total hip replacement by using a stainless steel femoral head and socket, but the artificial joint wore out after a short time. Sir John Charnley (1911–1982) of Manchester, England, ushered in the science of biomechanics by studying the forces and stresses of bones and joints and the body as a machine. After three decades of experimenting, in 1962 Charnley matched a stainless steel femur with a high-density polyethylene socket to create a prosthesis that would not wear down rapidly. Six years later, he developed a plastic cement, methyl methacrylate, to replace traditional metal screws.

Charnley's work in biomechanics heralded a new day in orthopedics. He referred to his work as *low-friction arthroplasty*, replicating the strength, mobility, and durability of joints. Over time many implants deteriorate, but in the meantime provide comfort and mobility that the implant recipients would not otherwise enjoy.

GAIT ANALYSIS

The union of art and science in biomechanics is gait analysis. Leonardo da Vinci (1452–1519) pioneered the study of bones and muscles by thinking of man as a machine. Galileo (1564–1642), Descartes (1596–1650), and Robert Hooke (1635–1703) also saw a relation between physics and the human body.

In gait analysis, diodes (simple electronic devices) are attached to fixed points on the body. The diodes send out pulses of infrared light to cameras photographing at least 315 frames per second. The cameras feed information from all the points into a computer that generates images pinpointing the location of stress on the joints. For example, the ball-and-socket joint of the hip bears six times the person's body weight each time he or she takes a step. Thus in a 150-pound person, each step would place 900 pounds of pressure on the joint. Most implants wear out under this amount of constant pressure.

Gait analysis applies to many other joints besides the hip. For example, whereas knee joint replacement was designed to meet the workings of the knee as a simple hinge, gait analysis has shown that these replacements fail because the knee moves all at one time along more than one axis. More than four hundred artificial knee joints have been designed, but none are perfect.

LOOKING INTO THE KNEE

Techniques of knee replacement have come a long way since William Hey's (1736–1819) simple description at the end of the eighteenth century of the "internal derangement of the knee." Yet his work ushered in useful theories of the anatomy and physiology of the knee.

The development of arthroscopy changed knee surgery. The word *arthroscopy* comes from the words *arthr*, meaning "joint," and *scope*, meaning "see." The origin of *endoscopy*, the science of looking into an organ or body structure, dates back to Botzinim (1773–1809), who in 1806 presented his *Lichterer* (meaning "light") at an academy in Vienna. Thomas Edison's (1847–1931) invention of the incandescent lamp made possible the use of light to view the dark recesses of the body. In 1918, Japanese physician Kangi Takagi (1888–1963) first used the endoscope to view a cadaver's knee. Later, he developed twelve different telescopes, a biopsy punch, and a flexible cauterizer for use with the arthroscope. This instrument enabled surgeons to operate without opening up and exposing the knee.

The arthroscope is like a tiny telescope with a half-dozen lenses that can be inserted into a quarter-inch incision. Optic fibers throw light in the pathway to illuminate the entire joint. The surgeon inserts tiny knives through the small incisions and removes only the cartilage that is damaged. Television technology was added to arthroscopy in 1975. Among the advantages of arthroscopy over open knee surgery are shorter hospital stays and reduced need for pain-killing drugs.

Physicians work to continually improve arthroscopy, developing new monitoring systems, sterilization units, and surgical equipment. Arthroscopy is not limited to the knee but is also useful in other joints, even on spinal disks.

ELECTRICITY IN BONE GROWTH

Electrical currents may help the mending of non-union fractures, or breaks that do not heal. About 5 percent of all fractures fail to heal normally. Luigi Galvani's (1737–1798) experiments in 1791 evoked the occasional use of electrical machines in London in the nineteenth century.

Modern interest stems from research in Japan in 1953. Yasuda Kinuko found that when bone is stressed or deformed, an electric charge is produced. This phenomenon is the *piezoelectric effect*. During the twentieth century, scientists studied electrical potential in mending non-union fractures. Studies of electricity on the behavior of fibroblasts (see Chapter 5) *in vitro*, or in a test tube in the laboratory, led to the idea that the same procedure might work in the body.

Andrew Bassett (1924–1994) and colleagues at Columbia University found that pulsed electromagnetic fields (PEMF) generate weak currents at a fracture to generate bone repair. Apparently, the fields promote synthesis of fibroblasts *in vitro*. According to Bassett, PEMF increases calcium ions. This method is noninvasive, unlike diode implants. Beyond its success in conditions where bone union is difficult to achieve, this technique may have implications for the management of fracture healing and many other uses.

LOOKING BELOW THE SKIN

The most common way to look under the skin involves the x-ray. This imaging device is most useful for the skeletal system in terms of identifying fractures and understanding bone structure. Bones appear white on an x-ray image because they have greater density than the surrounding tissues.

Another form of medical imaging involves *computerized axial tomography*, or CT scans. The idea originated in 1921 when the Dutch radiologist Ziedses des Plants (1902–1993) moved an x-ray tube and film during exposure to blur everything but one plane of interest. The purpose of tomography is to provide a means of imaging a single plane at one time. Tomography involves rotating both subject and film around horizontal axes. The patient lies on a special table and the sliding CT scanner passes over, taking pictures one small section at a time.

Another form of medical imaging, nuclear *magnetic resonance imaging*, or MRI, is based on the fact that the nuclei of some atoms absorb or emit radiation of a certain frequency if they are placed in a steady magnetic field. The MRI is used to diagnose many diseases of the skeletal system as well as tumors and cancers.

Ultrasound high-frequency sound waves that image body organs provide information about the structure and function of body tissues. The technology was founded in 1877 by a man named Strutt, twelve years before x-rays. Sonar produces detailed images even in densely packed structures. To build an image, the sound beam must be moved either by shifting the aim or position of the transducer, a device that transmits energy from one system to another, or by activating a succession of separate transducers. In the skeletal system, ultrasound has been used to accelerate tibial fracture healing through the application of low-intensity pulsed sounds. It has been used in the detection of full-thickness tears of the rotator cuff in the shoulder with an accuracy as high as 84 percent.

The twentieth century ended with significant technological advances and much promising research. Researchers' visions ushered in new avenues toward bone discovery that lead to even more beneficial applications in the twenty-first century. (See "Holistic Healers.")

Holistic Healers

From Hippocrates and Galen to the bonesetters of the British Isles, a group of healers emerged who believed in a holistic (total person) concept of health care, downplaying drugs and emphasizing prevention and health. Several groups of holistic doctors, including chiropractors, naturopathic practitioners, and acupuncturists, have unique, alternative approaches to medical treatment.

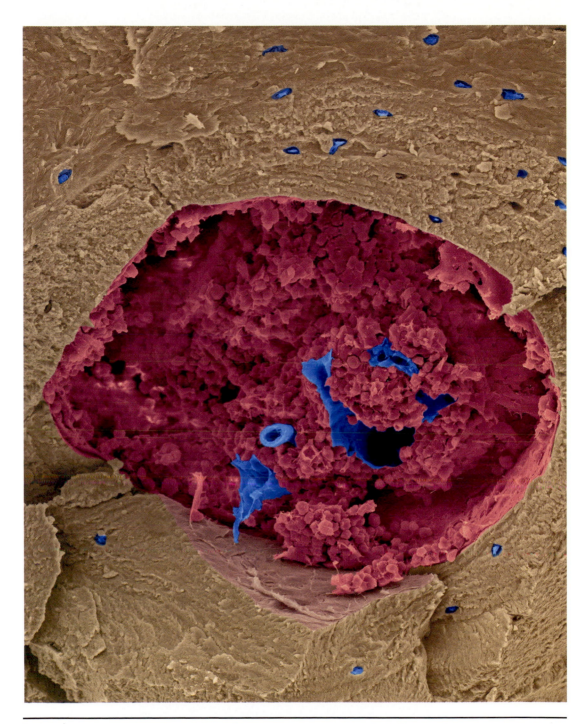

Scanning electron micrograph of bone and bone marrow cavity, magnified 70 times. Bone marrow appears as soft, spongy tissue in the middle of most larger bones. © Dennis Kunkel/Phototake.

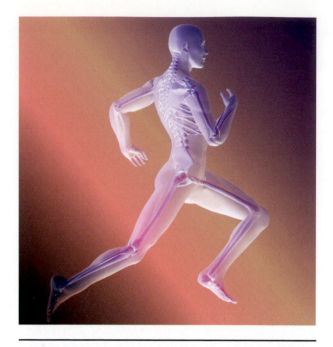

A figure of a runner showing the underlying skeleton. © Carol and Mike Werner/Phototake.

A thalidomide child painting at home, 1968. Getty Images/ Hulton Archive.

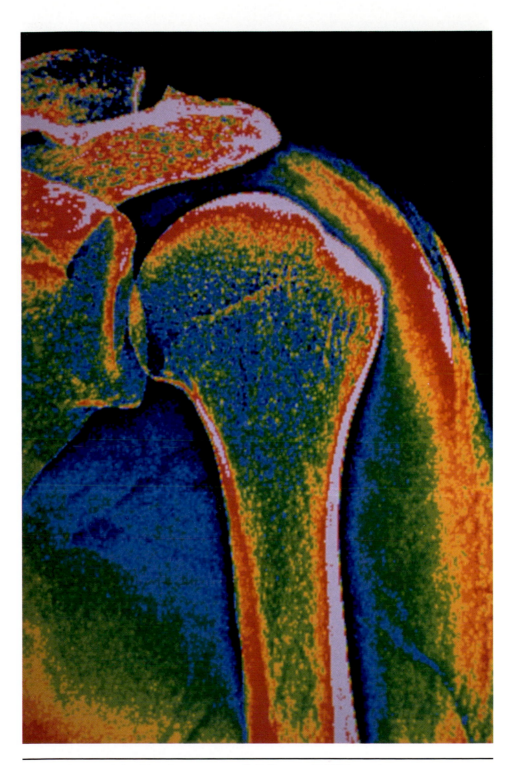

Colorized x-ray of shoulder, showing joints. © Collection CNRI/Phototake.

Photomicrograph of an osteon, or Haversian system, under polarized light. © James Hayden, RBP/Phototake.

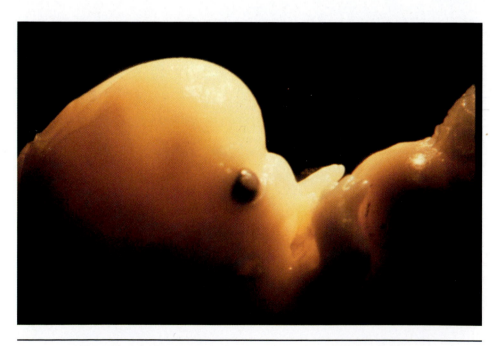

An image taken with a dissecting microscope at 6x of a normal human embryo (51–53 days gestation) showing part of the face, chest, and right arm. © R.A. Becker/Custom Medical Stock Photo.

Guy de Chauliac, "Twelve Surgical Instruments." A page from *Chirugia*, 1476. © National Library of Medicine.

Portrait of James Young Simpson, who invented and experimented with anesthesia. © National Library of Medicine.

Portrait of Hugh Owen Thomas. © National Library of Medicine.

Alexander Fleming. Portrait by Dean Fausett. © National Library of Medicine.

Female Ixodes Scapulous tick, a lyme disease vector. Centers for Disease Control and Prevention.

The Twenty-first Century and Beyond: A New World of Bones

The 1966 film *Fantastic Voyage* told the story of a mini-submarine whose mission was to dissolve a blood clot in the brain. Although this story was science fiction, it may not be far from realization in the twenty-first century. The promises of nanotechnology; tissue engineering, including stem cell research; and gene therapy will likely create a world in which scientists will know the exact mechanisms of bone and skeletal disorders, how to treat them, and eventually prevent them. By the end of the twenty-first century, research might end some of the world's most ancient medical problems, such as arthritis and bone deformities.

NANOTECHNOLOGY

With the tools of nanotechnology, scientists can print all 30,000 pages of the *Encyclopedia Britannica* on the head of a pin. The "nano" world, or world of 10^{-9}, has become one of the hottest topics of research in the twenty-first century.

Nano is a Greek word meaning "one-billionth" of something—for example, it could be a second, a liter, a meter. To avoid dealing with zeroes, scientists write it as 10^{-9}. The nanoworld involves very tiny things. For example, a nanometer equals about ten hydrogen atoms (hydrogen is the smallest atom) laid side by side, and one bone cell is several thousands of nanometers in diameter. (See "The Nanometer in Perspective" for other comparisons.)

The Nanometer in Perspective

- A nanometer is 1/1,000,000,000 of a meter.

- Individual atoms are less than a nanometer in diameter, measuring only a few angstroms (a few tenths of a nanometer) in diameter.

- Ten shoulder-to-shoulder hydrogen atoms equal one nanometer.

- One DNA molecule is 2.5 nanometers wide.

- One red blood cell measures thousands of nanometers in diameter.

- The head of a pin is one million nanometers in diameter.

- A 2-meter-tall man is two billion nanometers in height.

In 1905, Albert Einstein (1879–1955) was puzzled as he stirred sugar in a glass of water. As he watched it dissolve, he projected the size of one sugar molecule to be one nanometer in diameter. In 1959, Richard Feynman (1918–1988) gave a speech entitled "There's Plenty of Room at the Bottom." He asked the scientists in the audience, "What would happen if we could arrange atoms one by one in any way we want?" He suggested starting with atoms and building structures from them. With the challenge to begin at the bottom, several pioneers began to explore the world of nanotechnology.

In 1986, Eric Drexler penned *Engines of Creation: The Coming Era of Nanotechnology,* a futuristic book describing how this technology can be used to create a state of health and not just treat disease. He maintains that aging, illness, and injury are caused by scrambled or misarranged atoms—whether the disorder is caused by time, viruses, or swerving cars. If scientists can find out how to correct the misarranged atoms, Drexler argues, they can rebuild them one at a time.

In the nanoworld, the laws of physics are cast aside. Inertia, friction, and gravity do not matter. Particles behave differently from the same substances at normal size. Materials that never existed before can be built from the atom up. Researchers start at the bottom, working in the world of quantum mechanics (the theory in physics that assumes energy is not infinitely divisible).

Nanodevices and engineering efforts add to the developing fields of microelectronics, the technology that enables a computer to run. New microelectrical systems, or MEMS, currently in development will make today's fastest computer chip seem like a dinosaur. Structures like *fullerenes* or *bucky balls*, *carbon nanotubes*, *oxide nanobelts*, *nanowires*, *semi-conductor crystals*, *nanorobots*, and *quantum dots* will someday be familiar words. Nanotechnology will affect everything in the world—and out of it. It will lead to better and new consumer devices ranging from food, clothing, and cosmetics to plastics, cars, non-silicon-based computers, and spaceships. One scientist predicts self-cleaning windows and clothes, and even self-driving cars.

"Diamond Man"

Nanotechnology can produce by building—atom on atom—the hardest, most durable material of all: diamond. For those with bone loss or bone disease, nanorobots can deposit a diamond-like substructure that fills in skeletal deformities—as in brittle bone disease (see Chapter 18). Physician Robert Freitas (1957–   ), of the Zyvec Corporation in Richardson, Texas, predicts a whole world of medical robots that mimic clotting, white blood cells, and red blood cells.

Nanomedicine now involves hundreds of scientists working to develop effective drugs and materials that work at the cellular level. When surgeons now wield a knife they are working at the tissue level, but they are butchering individual cells at random in the process. Cell surgery, a new area of research, seeks to understand how changing chemical signals can cure conditions such as Paget's disease and osteoporosis (see Chapter 17). The cells are not damaged.

Whereas the prospects of healing a crushed or severed spinal cord with conventional medicine are bleak, with cell surgery it will become possible to remove scar tissue—which now interferes with healing—and guide cell growth to produce a healthy arrangement of cells on a microscopic scale. For fractures and injuries, cell surgery may direct cells to reorganize to make tissues and eventually re-grow entire organs and limbs.

Nowhere is there more hope for nanotechnology than in medicine, where it is important to see the body—including the skeletal system—from the perspective of its molecular makeup. Finding the secrets of the cells and their molecules is the challenge of the twenty-first century (see Table 13.1).

TISSUE ENGINEERING

When Theophilus Gluck cemented an ivory ball into a man's hip in 1871, he started a search for the right metal to help people move. Later innovators harvested bone grafts from the same person or other sources to replace bone. Now, a new group of researchers are leading medicine to novel pathways—looking not only to replace but actually to regenerate bone. The hope of the future resides in biology and cell-based therapy for orthopedic tissue

TABLE 13.1. Time Estimate for Nanotechnology Applications in Medicine and Pharmaceutical Products

2003–2010	2010–2015	2015–2025	Speculation
MEMS/nanodevices to separate DNA into sequences	Nanosensors that reduce medical errors by monitoring prescriptions from patients' homes	New tools for discovering drugs • Ink jet drugs • Molecular imprinting • Use of human cells as wave guides, assisting movement through tissue	Nanorobots to cure disease
A bio/silicon interface for diagnostics, sensors, pharmacogenetic drug discovery	Nanocomputing to allow faster drug discovery and modeling	Bio-inspired self-assembling nano-systems	Artificial immune systems
Fluid membrane networks for solid-state devices for microfluidics and microelectronics	New formulations for drug discovery	Vaccines and medicines based on an individual's genome	No more surgery
MEMS-based sensors for information on bone density and how grafts are progressing	New tools to characterize chemical/mechanical properties of cells	Synthetic molecular motors	
	Microscopic sensors to detect emerging disease		
	Biocompatible implants		
	Protein-folding models		
	Treatment for heart disease and diabetes using bioMEMs less than half the width of a human hair		

repair. The term *orthobiologics* refers to biological solutions to orthopedic problems. Tissue engineering to regenerate bone involves three elements:

- stem cells or other responsive cells
- growth factors
- a scaffold, or matrix, to support the formation of bone

Stem Cells

Stem cells in the embryo are cells that will eventually develop into all other cells. For years it has been known that mouse stem cells could be grown in the laboratory, but in the late 1990s James A. Thompson of the University of Wisconsin, Madison, found ways to make human embryonic cells grow. The procedure for developing stem cells is as follows:

- A human sperm fertilizes the egg.

- The fertilized egg grows into a clump of eight uniform cells, or the pre-embryo.

- The clump of cells becomes a blastocyst, with about one hundred cells creating an inner cell mass and a transparent shell that will be the future placenta.

- Cells from the inner mass are transferred to a culture dish and multiply with the help of special growth factors.

- An individual cell becomes multipotent (meaning "having many possibilities for development") and can renew itself, or differentiate, into many specialized cells such as bone, muscle, heart, and nerve. The cells can then be used in a variety of ways for treatment.

Although embryonic stem cells (ES) are the most effective in developing tissue, ethical questions cloud their use. Concerned with the federal limited ban on ES, researchers began looking at other sources for ES and located some in bone marrow, muscle, periosteum, and fat—with bone marrow being the most promising.

ADULT STEM CELLS

Just as in the early embryo, the bone marrow harbors cells that go through the process of developing. As these cells grow, those in the middle layer of cells (the mesoderm) and certain portions of the outer cell layers give rise to groups of stem cells that follow a similar sequence. These cells are called *human mesenchymal stem cells* (hMSC). They grow to increased numbers and then commit, or decide, what type of cell they will become—bone, cartilage, or tendon—and differentiate to form the tissue. Bone, for example, is formed by the assembling of a matrix, or network, that osteoblasts produce. In 1991, John Connolly reported on injecting marrow from a person as his or her own graft and concentrating the cells by using a centrifuge (a machine that separates substances by using the whirling motion of centrifugal force). Subsequent X-rays revealed that eighteen out of twenty tibial fractures treated with marrow adult cells had healed.

Refining these procedures, George Mueschler and a team from the Cleveland Clinic, Ohio, took out stem cells, concentrated them, and then grew them on a matrix. Mueschler is working with a micro-patterned polymer that increases the growth of cells more than 280 percent. Eventually, the researchers envision using nanodevices such as this "smart scaffold" (the polymer) and attaching MEMS-based sensors, which would emit wireless signals with information on bone density and the progress of healing. This discovery will benefit the 500,000 patients who undergo bone grafts each year and dread the bone marrow transplant more than spinal surgery.

GROWTH FACTORS

What sparks bone to grow in the body lies hidden in the deep recesses of dense skeletal fabric. Marshall Urist (1915–2001; see "Guru of Bone Growth), a researcher at University of California at Los Angeles (UCLA), found an important piece of the puzzle; **growth factors** called *bone morphogenetic protein* (BMP) and *transforming growth factor* (TGF). These were successful in healing animals and models and in trials. BMP triggers a charge that penetrates the cell and rearranges the basic chemical building blocks of DNA. However, using cells from humans evoked questions of purity, safety, virus inactivation, and biologic activity. Scientists at the University of Michigan have taken this research a step further by substituting a protein that programs for the gene for the growth factor. The gene, which remains active for a long time, has been effective in repairing long bone defects in rats and dogs.

Guru of Bone Growth

Arriving at UCLA from his native Belgium, Marshall Urist (1915–2001) fell in love with the desert. His habit of taking off two to three days in a row to explore the desert forced him to conduct his bone growth experiments quickly at cold temperatures to guard against bacteria growing while he was away. One day when he returned, he was amazed that his experiments standing over a period of time wielded the protein he had been looking for. He named the substance *bone morphogenetic protein*, or BMP.

As he worked to purify BMP, he was convinced this substance would revolutionize the surgical construction of bone. He made the protein into both powder and chips of demineralized bone that could be implanted to promote bone growth. When he implanted it in 135 patients with long bone defects, he reported healing in 80 percent of the patients.

Like other great scientists, Urist understood the importance of a natural occurrence such as bone metabolism long before it received widespread attention. His work fueled the search for improved fracture healing, bone induction, and bone conduction, or movement of substances to more remote points in the bone. A professor at UCLA for forty-six years, he received many national and international awards.

SCAFFOLDS

The third element for tissue engineering is a scaffold, or something to build the bone on. Many materials have been tried: calcium salts, ceramics, and coral. Currently, there is a boom in research for synthetic materials that avoid the risk of pathogens. Assembling complex proteins and sugars, scientists are using the same materials that already exist in the human body to make a scaffold for bone cells to grow on. Because the process resembles nature, scientists expect it to be well tolerated and not rejected. Engineering of tendons and ligaments is in its infancy; several scaffolds are in experimental phases but are a long way from clinical use.

In engineering tissue, it is important to understand not only the cellular mechanisms but also how structure relates to function. Just as simply having access to iron doesn't guarantee successful construction of a steel building, being able to manipulate cells doesn't guarantee successful tissue engineering. Enter the *bioengineer*, who works to understand the function in order to create the tissue.

After the cells and tissues have been grown, the structure must work. Nowhere is this more important than in the knee. Hanging free from the rest of the body, the knee is always bending, gliding, and rotating with the stresses that are put on it. In the leg, thirteen muscles support four major ligaments. Between the two leg bones are spongy, crescent-shaped menisci of cartilage. Hyaline cartilage does not have blood vessels or nerves and has limited potential for regeneration. Thus scientists are working with stem cells to regenerate cartilage. One lab is experimenting with stem cells in a light-sensitive gel; after it is injected, an invisible ultraviolet light hardens it. This becomes the scaffold on which the cell life is built. However, it must be able to handle the complex stresses of the knee.

GENE THERAPY

A map of the human genome is an awe-inspiring sight. Mapping the genome, or genetic make-up, of the human's twenty-three pairs of chromosomes constituted the work of thousands of scientists across the globe for more than fifteen years. It was announced as complete in 2000. The study reveals approximately 32,000 genes.

The 1998 *Oxford Medical Dictionary* defines *gene therapy* as "the treatment directed to curing genetic disease by introducing normal genes into patients to overcome the efforts of defective genes." The best target for gene therapy is one with a single gene. For example, it is known that Marfan syndrome (see Chapter 18) is caused by the presence of an abnormal gene that produces the protein fibrillin. If the abnormal gene could be replaced by a normal gene, then the condition might be reversed. The first gene therapy was tried in 1990 to treat persons with adenosine-deaminase, or ADA, deficiency, caused by absence of a single gene that programs for the enzyme ADA.

The use of gene therapy has been expanded to deliver specific proteins to tissues and cells. Understanding the role of proteins and how genes cause them to be expressed is the next thrust in molecular science; this field is called *proteomics*.

Proteins play a critical role in the regeneration of bone, cartilage, and ligaments. However, when a protein itself is put into the body, its life span is short. Gene therapy provides the gene that will program for the protein and enable the mechanism to be effective over the long run.

Gene therapy involves the following:

1. Choosing what protein to deliver

2. Choosing the correct target cell: osteoblast, myocyte, or fibroblast

3. Choosing a vector to deliver the protein; the vector is the means to get it to the cell

4. Choosing the way to get the vector to the cell: Directly inject it into the body; or indirectly remove the cell, expose it to the vector, and then reinsert (this is termed *ex vivo gene therapy*). Direct gene therapy has been used successfully in animal models to target severe diseases of joints, tendons, and ligaments.

In 2002, rheumatoid arthritis was the only orthopedic-related condition in a human gene therapy trial. The trial was conducted to establish safety and feasibility. At the time, no adverse reactions were reported.

PROTEOMICS

The science of proteomics involves cataloging and analyzing proteins in the human body. With the decoding of DNA in the human genome, the next step is to find the structure and function of proteins that are coded by the DNA. A segment of DNA is represented by four letters: A for adenine, T for thymine, C for cytosine, G for guanine. The sequence of these chemical compounds makes an **amino acid** chain, or protein, that codes for various substances. Explaining proteins is much more complex than explaining the place and position of the letters ATCG on the genome.

In 1995, the Australian researcher Marc Wilkins coined the term *proteome* and undertook the task of finding all the proteins in an effort similar to that of mapping the genome. The use of unique tools such as x-ray crystallography and nuclear magnetic resonance (NMR) have revealed the structures of many proteins. X-ray crystallography pictures reveal folds and kinks with different colors representing different functions. In the grooves and folds, certain molecules fit like a key in a lock. Scientists expect there are about one thousand shapes of proteins, such as barrel shapes, doughnuts, spheres, and zippers. Other advanced tools include crystallized protein scatter x-rays, which look for the functions of the folds and kinks; robots; and powerful x-ray generators.

Many researchers are studying the structure of proteins that encode for cartilage, tendons, ligaments, and bones. There are an estimated fifty thousand to two million proteins in the human body.

The future of nanotechnology, tissue engineering, stem cell research, gene therapy, and proteomics is promising. Although many of the ideas that are being advanced today will never reach clinical and practical application, it will only take proof of the effectiveness of a few to revolutionize the way skeletal diseases are treated and prevented.

The Injury-plagued Skeleton

For the children of America, thrills on the playground come with a price. One study revealed that 53 percent of all childhood fractures occur on the monkey bars, a disproportionately high rate when compared to injuries sustained on other playground equipment. The most common injury involves elbow fractures, which almost always require surgery. For children and adults alike, horseback riding leads to about seventy thousand visits to the emergency room each year, and that accounts for one injury in every two thousand hours of riding—three and a half times more than motorcycle-related injuries. These types of injuries are only three examples from accidents that occur because of carelessness, improper training, and use of improper equipment. Many others can occur during work, play, exercise, and sports.

SPRAINS

Although the word *sprain* is used to describe a variety of medical conditions, a true sprain is an injury to a ligament, the tough elastic band at a joint that connects bone to bone. Even though a ligament at any joint may be sprained, the most susceptible are knees, ankles, and fingers. (See photo.) A violent twist or stretch can damage the ligament and cause one of the three following types of sprain:

- *Simple stretch*: When ligament fibers become overstretched, pain is minor and little tenderness or swelling occurs. X-rays are generally normal, and the joint can still bear the weight of the person without much pain.

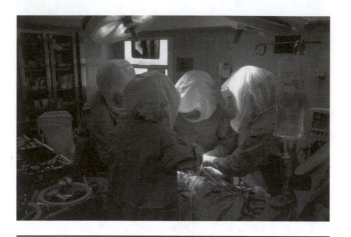

A team of orthopedic surgeons conducts a knee operation. ©
S. O'Brien/Custom Medical Stock Photo.

- *Partial tear.* The ligament fibers tear but do not rupture. Moving the joint is difficult and moderately painful. Some swelling and black-and-blue discoloration may be noted.

- *Complete tear.* This injury is the most severe. Joint function is impaired because the joint is completely misaligned. The area is swollen and appears black-and-blue. It may be necessary to immobilize the area with a cast or splint.

Any tumble or fall can injure a ligament, but recreational athletes most commonly suffer injuries to ligaments in the ankle and knee joints.

Sprained ankles occur when the tough fibrous ligaments that hold the anklebones in place are stressed. If a person falls in an abnormal position, the weight of the whole body is placed on those ligaments, causing them to stretch or tear. Ankle sprains usually occur on the outside, making the foot turn inward and causing excess tension on the outer ankle ligaments (see Figure 14.1).

Sprained knees occur most commonly when the anterior cruciate ligament (ACL) is twisted or rotated suddenly, as when a runner makes a sudden cutting movement to one side or the other. Likewise, a blow to the outer knee when the leg or foot is firmly planted on the ground or floor damages the medial collateral ligament (MCL), the large supporting ligament on the inside of the knee.

For treatment of such injuries, the R.I.C.E. acronym—Rest, Ice, Compression, Elevation—is useful (see "R.I.C.E."). Some charts add a P in front of the acronym for Protection, which involves two things:

- Warm-up and cool-down exercises. Many amateur athletes neglect these very important parts of a fitness program because they are too eager to jump into the activity and too tired after the physical demands. They cause themselves pain and injury by not warming up and cooling down.

- Use of devices that tape, brace, or wrap knees, ankles, wrists, and elbows.

"Train—Don't Sprain" must be the motto of the recreational athlete. In general, abdominal muscles and the muscles on the front of the arms and legs

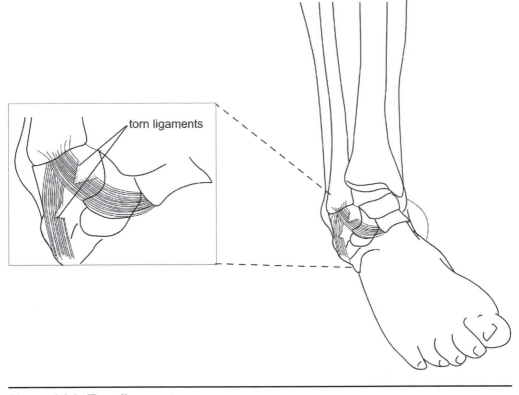

torn ligaments

Figure 14.1. Torn ligaments.

need strengthening. Back muscles and muscles on the back of the arms and legs need to be stretched before and after exercise (see photo).

DISLOCATIONS

A *dislocation* is a joint in which the bone is out of place. Generally, the misalignment is obvious. It usually occurs after a blow or accident with enough force to tear the ligaments. In addition, there may be damage to the joint capsule (the membrane that surrounds the joint), surrounding muscles, blood vessels, and bone.

Dislocations can be categorized as follows:

- Shoulder—This maneuverable, ball-and-socket joint is formed by the humerus and shoulder blade. Because the socket is shallow, it tends to be dislocated easily in one of four directions: forward, backward, up, or down.

- *Congenital*—These are dislocations that a person may be born with, such as a congenital hip dislocation or clubfoot (see Chapter 18).

R.I.C.E

When an injury occurs, it is important to remember the self-treatment acronym R.I.C.E. Following this safeguard can help speed recovery.

- **R—REST.** Rest is essential to healing; avoid putting pressure on the area, and curtail activity using the injured part.

- **I—ICE.** Apply ice in the form of packs, massage, or immersion. The cold slows metabolism within the cells, allowing tissue to survive a temporary lack of oxygen. Acting as a form of anesthesia, cold contracts vessels, controls bleeding, and relieves pain.

- **C—Compression.** Put an elastic bandage like an Ace bandage around the area until the swelling has subsided.

- **E—Elevation.** Raise the swollen arm or leg above the level of the heart to reduce swelling.

This treatment should begin as soon as possible within twenty-four hours of the injury.

"Photograph Illustrative of the Cololian Method of Treating Stiff Joints: Knee Exercises." Compare this device with the exercise machines of today. © National Library of Medicine.

- *Result of a complication*—A disease such as rheumatoid arthritis may cause a spontaneous dislocation, one not caused by injury.

- *Nursemaid's dislocation*—This occurs when an adult jerks or pulls a child's arm, causing an elbow dislocation.

- *Spinal vertebrae*—Dislocation to vertebrae can damage the spinal cord and cause paralysis.

- *Other*—Ankle, knee, wrist, or fingers can become dislocated by unusual stress. A lot depends on the joint's stability and the strength of the surrounding muscles and tendons. Shoulders and finger joints are not very stable. The hip is very stable but can be dislocated with a great deal of force, as when the knee hits the dashboard in a car accident.

With a dislocation, it is difficult or impossible to move the joint. Medical help should be sought as soon as possible. The physician will perform a *reduction*, the medical term for repositioning the joint, usually under general anesthesia. An untrained person should never attempt to put the bone back in place.

FRACTURES

A *fracture* is a broken bone. Although bones are rigid, because of their unique structure they can give somewhat when an outside force is applied. For example, if a person falls and lands on his or her outstretched hands, the force of the hands hitting the ground impacts the bones and connective tissue of the wrists. Usually these bones absorb the shock by bending slightly and then returning to their original position. However, if the force is too great or the bones have been weakened through malnutrition or conditions such as osteoporosis (see Chapter 17), the bones will break.

Children with broken bones heal much faster than adults. A bone that requires three to five months for the healing process in an adult will mend in four to six weeks in a child. Bone remodeling is very rapid in children—so rapid, in fact, that fractures may heal themselves so flawlessly that x-rays cannot show where the damage occurred.

A fracture may push the broken bone from its normal position; when the bone heals, there will be a condition called a *malunion*, meaning the bone parts did not join together properly. The bone parts now join at an angle instead of in a straight union, and the misformed joint is visible under the skin. If it causes no pain or problems, doctors will often leave the bone alone; however, if there is pain or other problems, they may have to re-break the bone and realign it. Malunion is more common in adults who do not get treatment promptly.

Kinds of Fractures

Several kinds of fractures are classified for diagnosis and treatment (see Figure 14.2) in the following ways:

- *Open or compound fracture*—The broken bone penetrates the skin and is visible. Considerable damage is done to bones and surrounding tissue.

- *Simple fracture*—A clean break occurs, and the skin is not broken. Neighboring muscles and other tissues are not damaged.

- *Greenstick fracture*—The bone cracks on one side only, not all the way through. This type of fracture usually occurs in children.

- *Comminuted fracture*—The bone breaks in more than two pieces or is crushed. This fracture may occur in an automobile accident or gunshot wound.

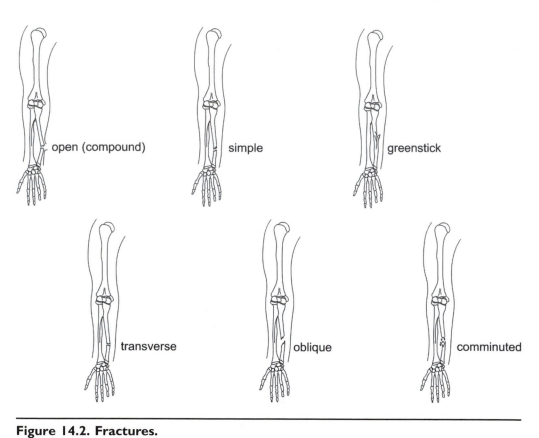

Figure 14.2. Fractures.

- *Impacted fracture*—The bone breaks, and then the pieces are forced into each other. This occurs in serious automobile accidents, such as head-on collisions.

- *Stress fracture*—A hairline crack, which is often invisible on x-rays for up to six weeks.

Recognizing a Fracture

In an open fracture, the exposed bone is evident and there is no doubt about the condition. But closed fractures are harder to recognize. Following are indications of a closed fracture:

1. A break or a snap is heard.

2. A grating sensation is felt with movement. This condition is called *crepitus*.

3. Swelling, bruising, or reddening may occur over the fracture.

4. The limb may look deformed—different in length, size, or shape from the other; it may be improperly angled.

5. The person may not be able to move the limb without intense pain.

6. Loss of sensation may occur at the end of the extremity; numbness or tingling may occur.

7. Other unusual pain may be present, such as in the rib cage when one takes a deep breath or coughs.

Breaking a bone is a shock to the entire body. Some people may feel dizzy or faint. Others may not feel pain at all because of the shock.

Falls and Hip Fractures among Older Adults

Among adults over 65, falls are the leading cause of fatal and non-fatal injuries as well as hospital admissions. The majority of fractures are caused by falls, the most common being broken vertebrae, hips, forearms, legs, ankles, pelvis, upper arms, and hands. Women sustain about 80 percent of all hip fractures due to osteoporosis (see Chapter 17). Most falls occur in and around the home, with common hazards including uneven flooring or carpet, lack of stair railings, slippery surfaces, unstable furniture, and poor lighting.

Skull Fractures

Head injuries cause more deaths and disabilities than any other neurological condition in people under age 50. Even with good care and treatment, mortality from head injuries is 50 percent or more. A *neurosurgeon* will treat skull fractures and any other brain conditions that result from a head injury.

Bleeding and visible bone fragments indicate a fractured skull. Other signs include discoloration behind the ear or around the eyes, bleeding from the nose, unequal size of the pupils, and obvious deformity such as swelling or depression. The word *concussion* comes from the Latin word *concussus*, meaning "shaken violently." A blow to the head or a fall may cause a concussion with temporary or prolonged unconsciousness. However, a person may have a concussion with or without a skull fracture.

Initial Care, or First Aid

For any suspected injury, it is important to stay calm and unhurried. However, it is also important to get professional trained help as soon as possible by calling an emergency number such as 911. The worst thing for a broken bone is to move it because serious damage could result. Also, an open wound is very serious because contamination from bacteria may enter the tissues and bone, causing infection. People giving first aid or medical treatment will splint, or brace, the injury and cover the open wound with a clean cloth or bandage. To reduce bleeding and swelling, it is important to elevate the wounded part.

At the emergency center, the physician will x-ray the fracture to determine the type and then will *reduce* the fracture. *Reduction* is the technical term for restoring the parts of the broken bones to their original positions. The guiding rule for all fractures is that broken pieces must be put back into position and prevented from moving until they are healed. Broken bone mends by knitting back together, with new bone forming around the edge of the broken parts.

Several factors influence the method of treatment:

1. The severity of the break

2. Whether the fracture is open or closed

3. What bone is broken (a spine or head injury is treated differently from a broken arm or rib)

Various types of treatment are used for fractures:

- *Cast immobilization.* Plaster and fiberglass casts are the most common types. The cast keeps the broken bone ends in place for healing.

- *Functional cast or brace.* This device allows for limited, controlled movement of a nearby joint. It may be desirable for some but not all fractures.

- *Traction.* A gentle, pulling action aligns the bone. Skin tapes or metal pins through the bone may transmit the force. Traction may be a preliminary treatment before other treatments are introduced.

- *Internal fixation or open reduction.* The orthopedic surgeon performs surgery on the bone. The fragments are reduced, or put into place, and held together with screws or metal plates. Rods may be inserted through the center of the bones. Because of the risk of infection and other complications, orthopedists use surgery only when they consider it the most likely procedure for optimal healing.

- *External fixation.* Pins or screws are placed in broken bones above or below the fracture site to reposition the bone fragments. These devices are connected to metal bars outside the skin and act as a stabilization frame to hold the bones in proper position.

Although a cast or fixation device is inconvenient and cumbersome, a broken bone cannot heal properly without it. Exercises during and after healing are essential to restore joint flexibility.

SPORTS INJURIES

Improving physical performance was a must for ancient man to stay alive to fight. Early Egyptian hieroglyphics show running, swimming, rowing, archery, and wrestling competitions. The first sports-related medical text was the Kahun Papyrus, dated about 1700 BCE. For the Olympic Games in 776 BCE, Greek athletes trained in gymnasiums under the supervision of specialists, or gymnasts. Thus sports medicine developed early in the history of man and ultimately became a major specialty within orthopedic surgery in the twentieth century.

Although the athlete may experience the thrill of success or the agony of defeat, other pains may occur: injuries. Anyone who regularly engages in sports runs a risk of injuring bones, joints, ligaments, or muscles. Many injuries require no treatment other than rest; others require immediate attention.

Shin Splints and Overuse Stress Injuries

Shin splints, known as *anterior compartment syndrome*, is one of the most common overuse problems. They occur as a result of the repeated straining of muscles between the shinbones, and they cause pain in the lower leg between the ankle and knee. In stress injuries muscles are fatigued, become swollen, and press on the blood vessels. Thus they do not absorb shock and cause muscles to transfer the stress to the bones, causing tiny cracks. Too much activity performed too rapidly, or the impact of a hard surface, or improper equipment increase physical stress. Shin splints may occur when a tennis player changes from a soft to a hard court or when a basketball player suddenly gets increased playing time.

The person first notices minor discomfort in the leg and realizes it hurts to touch one spot on the bone. If the symptoms are ignored, the condition may develop into a full-blown fracture and hurt all the time. Most stress fractures occur in weight-bearing bones in the lower leg and foot. Athletes in tennis, track and field, gymnastics, and basketball are very susceptible to this type of injury.

These microfractures are sometimes so fine that they do not show up on normal x-rays. Treatments include rest, maintaining a healthy diet with calcium-enriched foods, and use of proper equipment.

Women athletes experience more stress-related injuries than men athletes, probably as a result of the *female athlete triad*. This triad includes eating disorders (such as bulimia or anorexia as an attempt to maintain weight), infrequent menstrual cycles (because of excess dieting and hormonal changes due to heavy exercise), and early osteoporosis. As a woman's bone mass decreases, her chances of getting a stress fracture increase. In fact, 45 percent of competitive women runners develop stress fractures.

March Fractures

This type of injury may develop in one or more of the metatarsals as a result of repeated periods of excessive stress. Most common in walkers and runners, it causes pain on the ball of the foot. Resting and strapping the foot with adhesive plaster for a few weeks help in healing.

Knee Injuries

Acute knee injuries occur even though playing conditions have improved, stricter rules are enforced, and athletes are in better condition than ever before. The most common kind of injury involves the tearing of the meniscus, the band of cartilage supporting the knee. Knee position, direction, and magnitude of force determine tear type and location.

An anterior cruciate ligament (ACL) tear results from direct contact, as in football. Posterior cruciate ligament (PCL) injuries are most commonly the result of a posterior-directed force hitting the front of a bent knee. A medial collateral ligament (MCL) injury may result from an external rotation force, such as skiing or sustaining a straight lateral blow to the thigh or leg in football. (See Table 14.1.)

TABLE 14.1. Knee Injuries and Selected Sports

Baseball/softball	Dislocation of knee or trailing leg while taking an uncontrolled swing; 7 percent of all sliding injuries involve the knee.
Basketball	ACL, ligament, menisci injuries when another player falls on leg coming down from an off-balance jump; sudden stopping may cause tears; ACL injuries are common in women players.
Football	Most commonly injured joint in football; knees account for 23 to 36.5 percent of all football injuries; most common is MCL, followed by meniscal and ACL injuries; many are caused by tackling contact.
Soccer	Meniscal and ligament tears resulting from pivoting and deceleration rather than from contact, as in football.
Running	Fully 48 percent of all running injuries are knee injuries; common types are anterior knee pain, knee tendonitis.
Golf	Knee injury incidence is 9.3 percent among amateur golfers and 6.6 percent among professional golfers; meniscal tears are common.
Gymnastics	Knee and ankles are most affected joints; injuries occur during dismounts.
Racquet sports	Starting, stopping, cutting, and pivoting lead to injuries, similar to basketball and football injuries.
Volleyball	Knee tendinitis is the most common problem, caused by repetitive jumping and bending.
Ice skating	Knee problems are common; ligament and meniscal injuries and overuse injuries are frequent.

Hand and Wrist Injuries

Injuries to the bones and tendons of the hand are common in boxing, rock climbing, handball, and basketball. Although some people may take them lightly, broken fingers need medical attention. Fractures of the bones in the forearm are the most common in all groups. Usually they occur when a person falls on his or her outstretched hand. Children may have only a small amount of swelling and deformity, whereas adults may have a large amount of swelling.

The scaphoid is one of the small carpal bones in the wrist that is injured most exclusively in active young adults. It is a slow-healing fracture requiring three months in a cast that extends over the thumb. A severe fracture may require surgery with bone grafting.

Head Injuries

These are possible in most sports. A person may be knocked unconscious or may have a slight or serious concussion.

However, any blow to the head should be considered a medical emergency and treated to prevent permanent brain damage. The following signs may indicate a fracture:

- Bruising around the eye or ear
- Bleeding from the nose
- Pupils unequal in size
- Swelling of the skull

Most people are introduced to an orthopedic surgeon early in life. After all, each year more than a quarter of a million children under 5 years of age are treated at hospital emergency rooms for playground injuries. About 6.8 million people seek medical attention each year for accidental injuries, many of these involving bones. The skeletal system is strong and resilient, but the average person can expect to have at least two fractures over the course of a lifetime.

Diseases and Disorders
of the Joints

The English poet, John Milton (1608–1674), in his classic work *Paradise Lost* (Book XI), listed "joint-racking Rheuma" (rheumatism) as one of the terrible diseases that human beings must suffer because Eve had eaten the forbidden fruit in the Garden of Eden. Milton knew the curse that he was writing about, for he himself endured the agonizing pain of rheumatoid arthritis. As one of the oldest known conditions, arthritis is found in the skeletal remains of dinosaurs. Most people think Neanderthal man was born to stoop and shuffle, but samples of his perfectly preserved spine reveal that he was deformed with osteoarthritis. Moreover, as early as 8000 BCE, Egyptian mummies showed chronic arthritis of the spine.

Hippocrates (ca. 460–ca. 377 BCE) dubbed it *arthritis*, meaning "swollen joints," in 440 BCE, and Socrates (ca. 470–399 BCE) called it the most common disease of his time. The word *rheumatism* is derived from the Greek work *rheumatismos*, meaning "flowing or running of mucus." The Greeks thought mucus flowed from the brain to the joints and other parts of the body, producing pain. The Romans built special baths to treat people with arthritis, and the Emperor Diocletian exempted citizens most severely afflicted with the condition from paying taxes.

In more recent times, rheumatoid arthritis—the most crippling form—was documented in the 1880s and was thought to be the result of industrialization. However, bone studies of Native Americans of the American Southwest show the condition was present in that population around 1200. The Dutch artist Peter Paul Rubens (1577–1640) in the seventeenth century painted many individuals who might have had arthritis. Pierre Auguste Renoir (1841–1919), a French painter, is reputed to have strapped his brush to his deformed arthritic hands in order to paint. Throughout history, arthritis has been an "equal opportunity" disease, sparing neither young nor old, great nor small. Among its victims were Alexander the Great (356–323 BCE) Michelangelo (1475–1564), Isaac Newton (1642–1727), Benjamin Franklin (1706–1790), and Charles Darwin (1809–1882).

Actually, arthritis is not just one condition; it exists in one hundred or more forms. Examples are osteoarthritis, rheumatoid arthritis, and gout. The forms appear as different conditions but fit the definition of arthritis because inflammation of the joints is involved.

THE STRUCTURE OF JOINTS

Joints allow for relatively friction-free motion between bones. To understand the conditions that affect joints, a brief review of the structure of joints is necessary. Normal joint function depends on the effective operation of the following parts:

- A cushion of cartilage to absorb shock; this covers bone ends where two bones meet
- Synovial fluid to moisten and nourish the surfaces of the cartilage
- Cartilage, synovial fluid, and membrane to help joints move with little friction
- Outside the membrane, a joint capsule reinforced by ligaments that attach bones
- Fluid-filled sacs called *bursae* to separate muscles that cross bones or cross other muscles and to lubricate the movement of muscles
- Tendons that attach muscles to bones

When any one of these elements of the joint wears out or is affected by disease, arthritis can result.

In many kinds of arthritis, inflammation causes swelling, redness, heat, and pain. The damage from the swelling inside the joints may lead to deformity. Inflammation may be acute and limited, or chronic and persistent. According to the American Arthritis Association, there are ten divisions of arthritis:

- Degenerative joint disease, which includes osteoarthritis and osteoarthrosis
- Connective tissue disease, such as rheumatoid arthritis and juvenile arthritis
- Spondylitis, or arthritis of the spine

- Conditions associated with infectious agents
- Metabolic conditions, such as gout
- Neoplasms, or tumors, and **cancer**
- Conditions relating to nerves, such as carpal tunnel syndrome
- Disorders associated with articular manifestations
- Tendinitis and non-articular disorders
- Miscellaneous conditions that arise from other diseases, such as hemophilia

Nearly forty million Americans, or one in seven, have arthritis. (See photo.) One of the most prevalent chronic health problems, it limits everyday activities such as walking, dressing, and bathing. People in all age groups are affected, including 285,000 children.

OSTEOARTHRITIS (OA)

Osteoarthritis is one of the oldest and most common conditions known to man. Originally, wear and tear of the joint was considered to be the culprit, but scientists now realize that inflammation plays a major role. *Osteoarthrosis* is another name for the condition.

In OA, cartilage in the joints becomes thin, cracks, and breaks away, leaving bone unprotected. This loss of elasticity leaves bone ends unable to glide normally, and the stress damages the area, leading to inflammation, pain, stiffness, and limited movement. An obvious symptom may be a bony outgrowth, called *Heberden's* and *Bouchard's nodes*. The weight-bearing joints such as hips, knees, and spine are most often affected. However, if there is an injury, wrists, elbows, shoulders, ankles, or jaw may have symptoms.

Researchers think a number of factors contribute to OA: age, heredity, overuse of the joint, injury, obesity, and

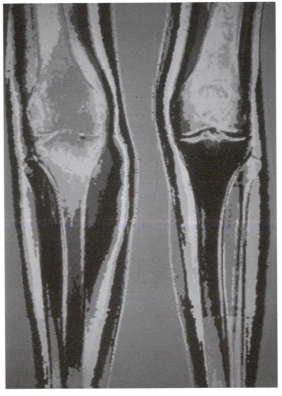

Front view of CT (computerized axial tomography) scans of legs, showing bone cancer on the right leg. The most common bone tumor is malignant myeloma, originating in the bone marrow. Other types include osteochondroma and osteogenic sarcoma. Collection CNRI/Phototake.

Corticosteroids: Use with Care

Corticosteroids are modified hormones from the adrenal gland. These medications have a strong anti-inflammatory effect and give physicians an option for use in conditions of arthritis that do not respond to other treatment. Injections of these drugs have been effective for acute bursitis and tendinitis and may relieve pain in joints affected by chronic osteoarthritis and rheumatoid arthritis.

There is a risk with the long-term use of these drugs. The side effects may include weight gain, puffiness, loss of bone mineral similar to that in osteoporosis, shrinkage of the adrenal glands, and other problems.

Corticosteroids should not be confused with *anabolic steroids* made from male hormones; athletes use anabolic steroids to increase muscle bulk. These anabolic steroids are exceedingly dangerous because of powerful side effects on organs such as the liver, heart, and reproductive systems. Corticosteroids are known by the names *cortisone* and *hydrocortisone*; the generic name is *prednisone*.

---

other diseases. This most common type of arthritis occurs mainly in people over age 54; approximately one-half of those over age 65 have OA. In 1992, researchers at the Boston University Arthritis Center confirmed in a major study that excess weight can induce or aggravate OA and that losing weight can reduce developing OA in the knees.

The recommended treatment for OA is aspirin; for those who have side effects with aspirin, nonaspirin pain products such as ibuprofen are effective. Nonsteroidal anti-inflammatory drugs (NSAIDs) are used to relieve symptoms in all forms of arthritis, and exercise is recommended to strengthen muscles and avoid stiffness. Physicians may also opt to prescribe **corticosteroids** (see "Corticosteroids: Use with Care").

RHEUMATOID ARTHRITIS (RA)

The immune system is generally considered to be a "good guy," fighting enemy invaders such as bacteria, viruses, and parasites. To combat these foreign bodies, the system must distinguish the "self," or its own cells, from the "non-self," or the outside invaders. When the immune system falters in this recognition, the B and T white blood cells that mediate immunity attack the self-proteins. This is what occurs in the host of diseases called *autoimmune diseases* (autoimmunity—when the body defenses attack normal tissue). Targets in RA are the person's own connective tissue, collagen, and synovial tissue around the joints.

Affecting some five to eight million people, RA is characterized by inflammation, pain, swollen joints (especially the small joints of hands), elbows, hips, and knees (see Figure 15.1). The course of the disease is unpredictable, and it varies greatly among people. Severe RA is active most of the time, lasts for years, and leads to joint damage and disability. Scar tissue may form between the bone ends and fuse the joint.

Other problems may accompany the disease, including anemia, inflamed eyes, pleurisy (pain on breathing), and recurring lumps of nodules at the back of the elbows.

In Search of a Cure

Because this mysterious disease has plagued humans for thousands of years, people have tried many cures. Suggestions have included the following:

- Sit in an abandoned radium mine.
- Bury oneself up to the neck in horse manure.
- Swig Dr. Finley's secret formula
- Chew Chuei-Fong-Tou-Gen-Wan
- Wear copper bracelets.

A young woman with rheumatoid arthritis using a specially designed spoon. © National Library of Medicine.

Years ago, people would try almost anything to get rid of the pain, so they were vulnerable to questionable products marketed as miracle cures.

Although many hoaxes still persist, scientists are making headway in unraveling the mysteries of the disease and providing relief from the pain and deformities. Because rheumatoid arthritis affects one in every one hundred Americans and can last for such a long time, beginning anywhere from age 20 to

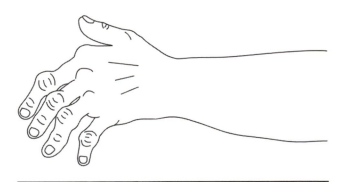

Figure 15.1. Hand showing rheumatoid arthritis.

50, research efforts are concentrated in this area. The following are areas of specialized research:

- *NSAIDs.* New NSAIDs (nonsteroidal anti-inflammatory drugs) are targeted at specific enzymes, chemical messengers known to be involved in RA. The enzyme cyclo-oxygenase-2, or COX-2, normally resides in the brain and other structures until inflammation causes it to rush to problem areas, such as joints. An *inhibitor* is a drug that stops the action of an enzyme. In 1990, COX-2 inhibitors (coxibs) were found to reduce inflammation and stop side effects. Introduced in 1999, celecoxib (Celebrex) is 400 times more effective than other NSAIDs. Another preparation, rofecoxib (Vioxx), is 1,000 times more selective for the COX-2 enzyme. The development of products that do not just relieve symptoms for RA has just begun; others are in the pipeline.

- *DMARDs.* Disease-modifying-anti-rheumatoid drugs (DMARDs) are preparations that seek not only to reduce pain and inflammation but to alter the course of the disease itself. The drug methotrexate, an **immunosuppressant**, was hailed as a major breakthrough in the 1990s. Other drugs are in development for people who do not respond to this treatment.

- *Biologic Response Modifiers (BRDs).* In November 1998, the U.S. Food and Drug Administration (FDA) approved the first of this new class of drug for use in reducing signs and symptoms of RA and inhibiting structural damage. The drug etanercept (Enbrel) heralded the entry of biotechnology into the RA marketplace.

- *Monoclonal Antibody (MAbs) technology.* Monoclonal antibodies (MAbs) are single antibody types produced from a B cell fused to a cancer cell. They can be made in large quantities in the laboratory with every molecule in the antibody mixture being identical. Because RA is an autoimmune disease, the use of MAbs is of interest because they bind to certain factors that are destroying joint tissue. Several MAbs have been approved for use with other drugs, such as methotrexate.

Researchers have found that thalidomide, the drug that caused the infamous medical disaster with "thalidomide babies" in the late 1950s (see Chapter 6), may have a safe form that controls rheumatoid arthritis and ankylosing spondylitis. (See "The Thalidomide Comeback.")

JUVENILE ARTHRITIS (JA)

Even though people do not expect to see arthritis in young adults or children, juvenile arthritis (JA) is a fact of life for 285,000 American children. To these young people, climbing stairs, throwing a ball, or opening a door is a challenge. JA is a form of rheumatoid arthritis, but its symptoms may be very different.

There are several types of JA:

The Thalidomide Comeback

Some researchers are considering thalidomide for treating a host of ailments that include rheumatoid arthritis, ankylosing spondylitis, and even AIDS. The form of thalidomide that had such terrible effects on newborns in the late 1950s (see Chapter 6) may have a *chiral* form (a chemical mirror image) that would serve as a beneficial treatment for some conditions. The inflammation of rheumatoid arthritis and ankylosing spondylitis may be caused by a factor called *tumor necrosis factor—alpha*, or TNF-alpha. When this factor is overproduced, inflammation occurs. Thalidomide blocks, or inhibits, TNF-alpha.

Thalidomide has been approved in the United States for use in patients who have Hansen's disease, or leprosy. One group of researchers has found that the drug can also treat the large growths in the mouths and throats of AIDS patients. Thalidomide may block the activity of certain genes that are necessary for HIV growth. The autoimmune disease scleroderma is another condition for which research with thalidomide is in trials.

- *Systemic JA* involves the entire body. The person may develop a high fever of up to 103 degrees and a rash. Pain in the joints follows the fever, as well as other inflammatory problems of the lining of the heart, stomach, and lungs. This type affects both boys and girls. It is the rarest kind of arthritis.

- *Polyarticular JA* occurs in many joints and resembles RA in adults. The small joints of the fingers and hands as well as weight-bearing joints are affected. About 40 to 50 percent of JA cases occur in this category. Girls tend to have it more than boys.

- *Pauciarticular JA* affects only a few joints. Boys tend to have low back stiffness. Girls may get eye inflammations that produce red eyes, eye pain, and failing vision. About 30 to 40 percent of children with JA have this type.

The causes of these types of JA are unknown. JA is not contagious, and heredity seems to play a part. Treatments include anti-inflammatory medications, rest, exercise, heat, and splints.

ANKYLOSING SPONDYLITIS

In the late 1960s, the writer Norman Cousins wrote the book *Anatomy of an Illness*, in which he tells of developing a painful condition called *ankylosing spondylitis*. Rather than denying the condition, he decided to defy it by laughing. So he rented comedies like *The Three Stooges* and found that 10 minutes of deep belly laughs could give him 2 painless hours of sleep. Using laughter, he cured himself of a condition that only one in five hundred people ever recover from. Because of ankylosing spondylitis,

Cousins started a very popular movement in medical circles of using humor in therapy.

The chronic condition that affected Cousins causes the vertebrae in the spine to fuse into a rigid mass. The inflammation may start outside the joints and then move to the hip and lower back, causing constant pain and stiffness. An abnormal curvature of the spine may cause a permanently stooped position. Because chest expansion is limited, lung and heart are affected. Only one-fifth of 1 percent of Americans have the disease; it usually affects men between the ages of 16 and 35.

Although the cause is unknown, scientists suspect a strong genetic, or family, link. Most people with the disease have the genetic marker HLA-B27. Genetic markers are protein molecules located on the surface of white blood cells that act as a type of name tag for finding a specific gene.

Treatment includes the use of NSAIDs to relieve pain and inflammation. Maintaining good posture is important, as well as following personalized exercise programs.

GOUT AND METABOLIC ARTHRITIS

Gout is a metabolic condition that occurs when something goes awry in the use of certain body chemicals in other parts of the body. For some reason, uric acid—which is derived from purines, food substances in the diet—begin to accumulate and are deposited as needle-like crystals in joints and other tissues. The big toe may suddenly become hot, swollen, and painful and appear red or purple and shiny. The condition may be genetic. It seems to be triggered by excessive drinking and eating, surgery, crash diets, trauma, or diuretics.

About one in one hundred Americans are affected. As many as 75 to 80 percent are men, with the first attack starting between 40 and 50 years of age. This is one arthritic condition that seems to respond quickly to medications. Colchicine and endomethasin are most often used. These agents increase the body's ability to lower uric acid levels in the blood.

A condition called *pseudogout*, or false gout, occurs when crystal deposits of a calcium salt affect the knees, wrists, and ankles. This condition strikes both men and women in equal numbers, usually in people over 70 years of age.

Other metabolic diseases that affect the joints include the following:

- *Vitamin C deficiency, or scurvy.* This disease gained recognition centuries ago when British sailors who did not eat fresh fruits for a long period of time developed extreme weakness in the joints. The Scotttish doctor James Lind (1716–1794) found that lime juice, which is high in vitamin C, prevented and cured this disease. The sailors were thus nicknamed "Limeys."

- *Ehlers-Danlos disease*. This inherited condition of the elastic connective tissue causes the person to develop a soft, velvety skin that bruises easily (see Chapter 18).

- *Marfan syndrome*. This inherited condition produces joint pain and other serious symptoms (see Chapter 18).

OTHER AUTOIMMUNE DISEASES

These diseases are similar to RA in that they involve the immune system turning on itself. A new approach in autoimmune research is to look not just for the underlying causes of a specific disease but also for common genetic pathways that cause groups of autoimmune conditions. (See "New Approach to Studying Arthritis.")

Lupus (Systemic Lupus Erythematosis)

The term *lupus*, meaning "wolf," is applied to this condition because the individual may develop a wolf-like red mask on the face. *Erythematosis* comes from the Greek word meaning "red."

This autoimmune disease attacks joints and other linings, producing joint inflammation. The person becomes very tired and develops a rash across the cheeks and nose that worsens with exposure to sunlight. Symptoms are unpredictable and may come and go. Corticosteroids and other medications are used to treat this disease. The typical person with lupus is a woman be-

New Approach to Studying Arthritis

Most arthritic joint conditions involve autoimmune diseases. When two German scientists proposed the idea over a hundred years ago, people thought it was preposterous that the body's immune system could turn against itself. The search for the meaning of autoimmunity led to the investigation of many diseases that had a great variety of symptoms and responses to treatments.

In January 2003, the National Institutes of Health announced a new priority of seeking common denominators among certain autoimmune diseases. The Multiple Autoimmune Diseases Genetics Consortium (MADGC) will attempt to identify genes that several autoimmune diseases have in common. The goal is to recruit as many members of a family as possible to locate those who have autoimmune diseases. The study will enroll four hundred pairs of twins and siblings who have rheumatoid arthritis, juvenile arthritis, and systemic lupus erythematosis. Parents and other family members will also be enrolled for genetic analysis. By following all study subjects for five years, the researchers hope to pinpoint genetic as well as environmental factors that affect the diseases.

tween the ages of 15 and 35. The disease sometimes appears during pregnancy and may cause one in four infants to be born prematurely.

Scleroderma

Literally meaning "hard skin," *scleroderma* is a progressive autoimmune
disease that can cause thickening and tightening of the skin as well as serious damage to internal organs, including lungs, heart, kidneys, and gas-
trointestinal tract. The severity of the condition varies; it can be just a
nuisance or a life-threatening illness. There are two basic types:

- *Systemic scleroderma* occurs when the immune system damages small
  blood vessels and collagen-producing cells in the skin and elsewhere
  throughout the body. Because the blood vessels are affected, the person may
  develop Raynard's phenomenon: blue color changes in the fingers when exposed to cold.
- *Localized scleroderma* affects collagen-producing cells in some skin areas
  but not in internal organs.

With the condition, women outnumber men four to one, and diagnosis is
usually made when people are in their forties. About 100,000 people have
scleroderma. Although no cure exists, researchers in 2003 think a cure
might be just around the corner. Investigators have identified a novel cell
death pathway, or the biochemical changes that cause the cells to die, in
scleroderma tissue that changes the structure of the body's molecules so
they recognize the changed tissues as foreign and attack them. A next step
is to develop medications to interfere with the cell death pathway. Several
medications control some symptoms, including that of Raynard's phenomenon, by causing the blood vessels to dilate, or open up.

Polymyocitis

This rare condition is characterized by inflammation of the muscles, causing pain, swelling, heat, and redness in the small joints and reddish patches
on the skin of face, knuckles, elbows, knees, and ankles. Women ages 30 to
60 are twice as likely to have the condition as men. Children between the
ages of 5 and 15 may develop the disease.

Infectious Arthritis

Many bacteria can invade joints and cause acute or chronic arthritis. The
bacteria in the gonococcus that causes gonorrhea, a sexually transmitted disease, and the meningococcus that causes meningitis are two examples of infections that may cause joint inflammation. For many years, tuberculosis
(TB) of the spine was treated as a separate disease until a researcher realized the bacterium *Mycobacterium,* which causes TB, also causes infectious

arthritis. Both diseases involve infection of the synovial fluid and produce painful swelling, redness, and heat in the joints. Likewise, certain viruses, fungi, and parasites may cause infectious arthritis.

Bursitis

More than 160 tiny, fluid-filled sacs are packed into spaces where muscles or tendons move over bones or other muscles. These little sacs, called *bursae* from the Greek word meaning "wineskins," cushion the muscles and bones as well as other parts of the joints so they can move freely and without friction (see Figure 15.2). But when an area sustains a sharp blow, excessive pressure, or overuse, the little sacs can become inflamed, creating a condition known as *bursitis*. Vulnerable areas are: shoulder, knee, heel, elbow, hip, and groin. Treatment includes rest, use of medication such as NSAIDs, and injections with corticosteroids.

Lyme Disease (LD)

When a cluster of children around Lyme, Connecticut, came down with arthritis in 1977, the hunt for the culprit led them to the deer tick, *Ixodes scapularii* (see photo in color section). Also known as the black-legged tick,

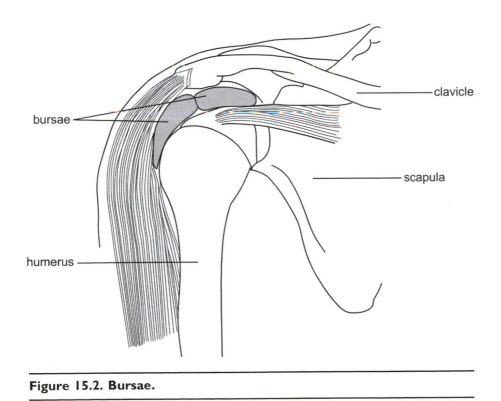

Figure 15.2. Bursae.

the small parasite harbors the bacterium *Borrelia burdorferi*, an organism that causes a condition with a number of symptoms, including arthritis. Most of the ticks are located in the northeastern and north-central United States. A western variety, *Ixodes pacificus*, may transmit the disease along the Pacific Coast. The risk for LD is mostly localized to these areas.

Compared to the common tick, found on cats and dogs, *Ixodes* ticks are tiny. In the larval and nymph stages, they are no bigger than a poppy seed. The adult stage is about the size of a sesame seed. Ticks feed on the host's blood by inserting their mouths into the host's skin. In the process of feeding, they transmit the bacterium after attaching themselves for two or more days.

The early symptoms of LD are mild and easily overlooked. The first symptom is an expanding rash called *erythema migrans*. Some people describe the rash as a bull's-eye because it has a central white spot (where the tick has attached) surrounded by a solid red, expanding rash. The blotch has an average diameter of 5 to 6 inches but may extend to 2 feet. The rash usually persists for about 3 to 5 weeks and is accompanied by joint pain, chills, fever, and fatigue. Later on, other symptoms develop, including severe headaches, painful arthritis, heart problems, central nervous system disorders, and possibly mental problems like memory loss and disorientation. The condition is often misdiagnosed as arthritis due to these symptoms, but, unfortunately, the patient will not respond to arthritis treatments.

Diagnosis of LD is a subject of considerable debate. At the moment, there are at least five FDA-approved tests. Assuming prompt diagnosis, early treatment (begun in the first three weeks) usually ends in a full cure. Three oral antibiotics—doxycycline, amoxicillin, and ceften—are recommended for all but a few symptoms. However, response to treatment of late Lyme disease varies greatly; the multiple use of oral and injected antibiotics may cause adverse side effects and long-term autoimmune conditions.

REPETITIVE MOVEMENT DISORDERS

Tendons connect muscles to bones; cartilage connects bone to bone. In many parts of the body, tendons are interwoven with muscles or serve as short connections between the ends of muscles and bones. Throughout the hands and feet, tendons are like long cords. Fibrous tendons promote movement among muscles and bones. Repetitive motion of the joints affects the structures that hold the bones together causing a variety of problems. These conditions are called repetitive movement disorders.

Trigger Finger and Tennis Elbow

Trigger finger is an example of tenosynovitis, an inflammation of the protective sheath around the tendon called the *synovium*. In this condition, the finger clicks and pops as it is bent and straightened. Eventually, the person may not be able to straighten the finger. Overuse of repetitive motions, such as those required by typists or factory workers, may cause the condition. A bacterial infection from a puncture wound may damage the area also. Tenosynovitis can occur in the wrist or in the shoulder (the bicipital tendon). Medication usually relieves the symptoms, but sometimes surgery is necessary.

Tennis elbow is an example of tendinitis, or inflammation of a tendon. The painful symptoms occur just outside a joint, especially the elbow or shoulder. The cause is a small tear in or inflammation of the tendon that links muscle to bone. A chronic condition may result if the affected area is not rested. In older people who continue to use the joint, the tendons heal slowly and the ligaments and tendons gradually stiffen, causing a "frozen shoulder." Rest and medication for pain may help, but reconstructive surgery is usually necessary.

Writer's Cramp and Washerwoman's Thumb: Carpal Tunnel Syndrome (CTS)

An alternating numbness and tingling sensation runs through the fingers and hands. They feel like they are asleep. Pain then shoots from the wrist up into the forearm or down into the palm. The person cannot sleep at night. Relief can only be gained by getting up and walking around and shaking the hands. These are the symptoms of carpal tunnel syndrome (CTS).

Named from the Greek word *karpos*, meaning "wrist," the carpal tunnel is a passageway through the wrist. Bordered by bones and ligaments, it shields the nerves and tendons that extend into the hand. As the fingers move, the tendons that control them slide back and forth beside the bones and ligaments that form the tunnel. But when the tissues around the tunnel swell and place pressure on the median nerve that provides feeling to the thumb, index, middle, and ring fingers, pain results. The little finger is spared; this fact is sometimes an indication of CTS.

CTS is associated with repetitive motions in the workplace, but anyone who places a lot of stress on the wrist may get it. The most stressful actions are wringing movements of the hand associated with playing a musical instrument, using power tools that vibrate, typing, or tightly grasping a steering wheel. The motion of wringing clothes has led to its nickname as *washerwoman's thumb*. Gripping or pinching with the fingers while the wrist is flexed may stress the area. For example, blacksmiths have a high incidence of CTS. Also at risk are carpenters, grocery clerks, meat cutters, mechanics, and sports enthusiasts such as golfers and canoeists. (See "Preventing CTS.")

If CTS is suspected, one should see a physician, who will perform tests to assess damage. A nerve conduction velocity test pinpoints how much the median nerve is pinched. Treatment involves mobilizing the hand with a light-weight plastic splint that still allows some motion and use. Corticosteroids may relieve the inflammation, but surgery may be necessary.

Preventing CTS

Anyone who uses his or her hands for repetitive movements should be aware of the risks of carpal tunnel syndrome. The following guidelines help prevent this painful syndrome:

- Take short breaks from hand-intense work every half hour or so even if there is no pain. Move the hands around.

- Do not lift objects without using the thumb and index finger; use the entire hand and all the fingers.

- Do not hold the hands in the same position or keep the wrists flexed for long periods.

- Do not sleep on the hands.

- When typing, use a light stroke and do not rest the wrists on the keyboard or desk. Keep fingers lower than wrists.

- Do not choke the steering wheel when driving; hold it gently (but firmly enough to maintain control, of course).

OTHER JOINT DISEASES OF YOUNG PEOPLE

Perthes Disease, or Legg-Calve-Perthes Disease (LCPD)

Perthes disease is a disorder of the hip in young children. Also called Legg-Calve-Perthes disease (LCPD), the condition was described independently in 1910 by Arthur Legg (1874–1939), an American; Jacques Calve (1875–1954), a Frenchman; and Georg Perthes (1869–1927), a German. The disease occurs in children from about ages 4 to 10 and is more prevalent in boys than in girls.

Perthes is characterized by loss of circulation to the head of the femur where it fits into the socket, or acetabulum, of the hip. This results in avascular **necrosis**, the death of bone cells in the area. Due to the loss of blood, the area may become brittle and collapse, leading to deformity and arthritis.

In the growing child, the femoral head has several distinct zones:

- *Metaphysis*, or spongy area, next to the ball
- *Physis*, or growth center, composed of bone-forming cells
- *Epiphysis*, located at the ball end; this part fits into the acetabulum and enables the bone to grow

For unknown reasons, the circulation in Perthes disease is cut off to one or more parts of the femoral head, causing the cell death.

Symptoms vary according to the stage of the disease. In the first stage, or **ischemia** state, in which the blood supply is temporarily cut off, the child may complain of a painful hip due to the inflammation of the synovial lining of the joint. Occasionally, the child may feel pain in the knee. The second stage occurs when the blood supply has been fully shut off and the body attempts to repair the damage. The next stage, called *revascularization*, occurs when the blood flow is returned. During this stage, a fracture of the head of the femur may occur. A parent may notice that the child has a sudden limp that lasts for days or weeks. In the second stage, the bone goes through repair and remodeling. It is not uncommon for a child to go through the entire process of ischemia, revascularization, fracture and collapse, and repair and remodeling with no significant symptoms.

When a child has acute pain in the hip accompanied by a limp, the physician will usually order an x-ray, which in the ischemia stage may not reveal the condition. Blood tests may be ordered to rule out infection or other acute problems. Several conditions mimic Perthes disease, including hypothyroidism (lack of thyroid hormone) and epiphyseal dysplasia (a disease of the growth plates of the body). Occasionally, the doctor will order a bone scan to confirm the diagnosis.

Once the doctor is satisfied the problems are caused by Perthes disease, a course of treatment must be determined. Certain factors influence this decision:

- If the onset of the disease is before age 4, results tend to be good with or without treatment. The older the child, the less likely the child will improve without treatment.

- If the hip is stiff and the leg cannot be moved outward from midline, results are not favorable. The patient may be placed on traction, a process to draw or pull the limb, to relieve the stiffness or be given physical therapy. Surgery may be necessary to correct the tightness.

- If x-rays reveal partial or total hip involvement or involvement of the whole head of the femur, the prognosis is not favorable.

In the past children were placed on prolonged bed rest, given crutches, and encouraged to put no weight on the area. This approach is no longer valid. Today, physicians follow the policy of *containment*, which involves casting, bracing, or performing surgery. The casts are usually worn for eighteen to twenty-four months. If surgery is necessary, the femur is cut just below the hip and the head of the femur is redirected into the acetabulum. Plates and screws are inserted to hold the fixture in place.

As many as 60 to 70 percent of children with Perthes disease do well with no long-term disability. Controversy remains over whether or not to treat and the best type of treatment. Each case has unique factors that influence the ultimate decision.

Slipped Capital Femoral Epiphysis (SCFE)

Slipped capital femoral epiphysis, the most common hip disorder among young teenagers, starts when part of the growing end (epiphysis) of the thighbone (femur) slips out of place. In addition to pain in the knee or thigh, other problems may occur: The person may not be able to turn the hip inward, the foot may turn outward, and one leg may be measurably shorter than the other. SCFE can develop in one leg or both legs.

The condition usually occurs in children between the ages of 11 and 16, often in children who are overweight. More boys than girls get SCFE. The cause is unknown. Two types of SCFE are seen:

- Stable SCFE. More than 90 percent of cases fit into this category. The child can walk with or without crutches, and the stiffness in the hip may get better after a period of rest. Later a limp appears, and the pain comes and goes. Sometimes pain is not felt directly in the hip but in the groin, thigh, or knee.

- Unstable SCFE. The pain is excruciating, like that of a broken bone, and the person may not be able to move the leg.

To make the diagnosis, the doctor will order x-rays of the pelvis or thigh areas from different angles. Other tests may then be performed. With this condition, surgery is the treatment of choice. The orthopedic surgeon will perform an *in situ* fixation in which a single central screw is inserted to hold the bone in place. The screw keeps the femur from slipping and enables the growth plate to close.

After the surgery, it takes four to six weeks to get back to normal. The child then can slowly resume regular activities, including running and contact sports. If the disorder is treated early, chances of a full recovery are good. However, some complications may result. The most serious are avascular necrosis, or lack of blood flow to the bone that causes cell death, and chondrolysis, or decay of cartilage. Necrosis is more common in people with unstable SCFE. Some children who have had SCFE may develop arthritis in the hip in later life.

Osgood-Schlatter Disease (OSD)

Osgood-Schlatter disease is one of the most common causes of knee pain in young people. Named after two physicians—the American orthopedic surgeon Robert B. Osgood (1873–1956) and the Swiss surgeon Carl Schlatter (1864–1934)—the condition is more a problem of overuse than a disease.

Young athletes, especially, get OSD during their rapid growth years, ages 9 to 13. It is more common in boys than in girls.

OSD results when the large powerful muscles in front of the thigh (quadriceps) pull on the patellar tendon. As a result of overuse, these tendons start to pull away from the bone. Usually the problem affects only one knee, which may be swollen, warm, and tender to the touch; the disease appears as a bony bump below the kneecap. When pressed, the bump hurts. It is also painful when the young person kneels, jumps, climbs stairs, runs, squats, or lifts weights.

The doctor may want to x-ray the knee to make sure the problem is not caused by another condition. If OSD is diagnosed, the young person may be advised to cut down on activity and avoid any deep knee bending until the pain has been gone for two to four months. The doctor may recommend wrapping the knee in an elastic bandage and implementing R.I.C.E.—rest, ice, compression, elevation (see Chapter 14). OSD usually goes away after the person stops growing.

RELATED CONDITIONS

Costochondritis and Tietze Syndrome

Costochrondritis is an inflammation of the cartilage that connects the inner end of each rib with the breastbone. This painful condition causes swelling in the front of the chest where the ribs join the sternum. The word comes from the Latin *costo*, meaning "rib," and *chondro*, meaning "cartilage." It is most common in young adults but can occur in any age group. In the United States, the condition accounts for 10 percent of chest pain episodes in the general population and 30 percent of chest pain episodes seen in emergency rooms.

Although inflammation or a blow may cause this condition, often the cause is unknown. Because pain occurs in the chest, it may be confused with a heart attack. The pain is usually located around the second and third ribs and is often related to movement, coughing, or sneezing. Because the symptoms are so similar to those of a heart attack, it is wise never to take them lightly. Given time and rest, the symptoms usually go away.

Costochondritis should be distinguished from Teitze syndrome, which involves the same area of the chest. In this condition, swelling is common, it indicates inflammation of the costochondral cartilage in the upper front of the chest. Redness, tenderness, and heat are present. The sharp pain is often confused with that of a heart attack. Once the syndrome is diagnosed, the same treatment is used as for costochondritis and Teitze syndrome.

Popular Names for Joint Disorders

Many people have a form of arthritis without knowing it. Interesting names describe various conditions (see Figure 15.3). Here are some of the most interesting:

- *Tailor's seat* or *weaver's bottom* describes an inflammation in the bursae over the bone in the posterior caused by sitting too long.

- *Tennis elbow* just doesn't happen from playing tennis but can be caused by fly casting, frisbee throwing, rug beating, or any other sudden tightening of the muscles in the hand or forearm.

- *Housemaid's knee* is bursitis in front of the kneecap. This is different from *nun's knee* or *clergyman's knee*, where bursitis occurs below the kneecap.

- *Student's elbow* arises from habitually resting the elbow on a desk.

- *Video-game finger* is caused by the repeated motion of playing video games.

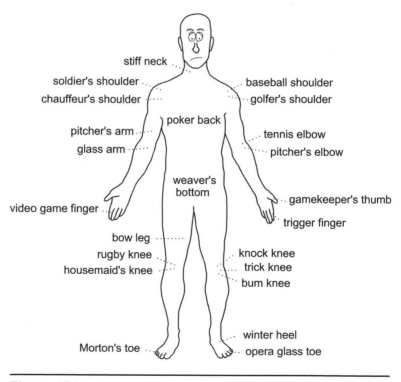

Figure 15.3. Arthritis.

Osteonecrosis of the Hip

Osteonecrosis of the hip is a severely disabling condition that can lead to the collapse of the hip joint. The symptoms are similar to those of Perthes disease, although the two are not related. In osteonecrosis of the hip, the blood vessels slowly cut off nourishment to the top of the femur (thighbone) where it fits into the socket of the hip (acetabulum). Without the blood, the head of the thighbone dies and collapses. The result is a very painful hip condition. The cartilage in the socket may also break down. The condition usually starts in one side of the hip and eventually goes to the other.

About ten thousand to twenty thousand cases of osteonecrosis are diagnosed each year. Risk factors include: age 20 to 50; hip dislocation or fracture; alcoholism; use of corticosteroids; glandular problems; other diseases such as rheumatoid arthritis.

An x-ray and MRI (magnetic resonance imaging) will reveal if the damage is only to the head of the thighbone or if the bone marrow is dying or dead. Depending on the diagnosis, the following options may be presented:

- If the head of the femur is not yet collapsed, the physician may use bone grafting or decompression, or removal of pressure, to help the body build new blood vessels and bone cells.

- If the hip has already collapsed, total hip replacement (arthroplasty) may eliminate pain and promote mobility. In this procedure, a ball and socket replace the hip joint. The thighbone is fitted with the ball, which fits into the socket. Diseases and disorders of the joints are ancient afflictions that are beginning to be understood. The goal of researchers is to stamp out or control these problems. Recent breakthroughs indicate that target may be in sight.

Feet, Hands, and Backs

Only in humans are the roles of the hands and feet separate. Other primates walk on all fours and use their feet as hands. Many disorders can affect the hands and feet—some of them very simple, others quite serious. This chapter discusses a range of potential problems of the hands, feet, and back.

FOCUS ON FEET

According to the American Podiatric Medical Association, the average person will walk about 115,000 miles during a lifetime—more than four times the distance around the equator. Thus it is no wonder that words like *corns*, *bunions*, and *ingrown toenails* are so common.

Leonardo da Vinci (1452–1519) called the feet a masterpiece of engineering and a work of art. Fifty-two bones, 35 joints, and over 100 ligaments make up the feet. Bones in the feet constitute one-quarter of the body's 206 bones. The heel, the base of the little toe, and the base of the big toe form a weight-bearing tripod. Toes are added for balance. During any given day a person takes 6,800 to 10,000 steps on tile, pavement, and other surfaces. With each step, gravity-induced pressure of three to four times the body's weight bears down on each foot. An examination of the bones and appearance of the feet can shed understanding on problems such as arthritis, diabetes, circulatory disease, and the side effects of medicine.

Bunions

Bunions are bony protrusions under the skin at the base of the big toe, making the big toe bend toward the second toe and sometimes overlap it.

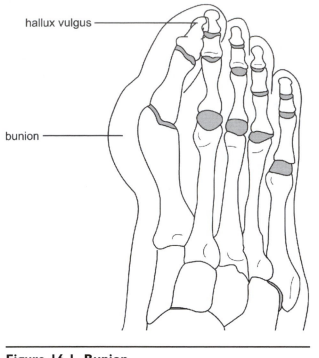

hallux vulgus

bunion

Figure 16.1. Bunion.

The condition, known as *hallux vulgus*, is named from the Latin *hallus*, for "big toe," and *vulgus*, referring to deformity (see Figure 16.1).

More women than men develop bunions. Wearing high-heeled shoes with pointed toes forces the foot to re-form. There also may be a genetic component. Although the bunion itself is a mild discomfort, it may be accompanied by osteoarthritis or bursitis. The problem increases with age and is aggravated by obesity because the pad of fat under the heel flattens and becomes less resilient. Thus stress is placed on the muscles and bones of the feet when walking or standing. Shoes must fit properly to provide ample room for the toes. To correct bunion problems, people should consult an orthopedic surgeon or a podiatrist, a doctor who specializes in foot care. Occasionally, surgery may be necessary to align the bones. The common surgical procedure involves removing some cartilage and bone and remodeling the deformed joint.

Flat Feet

Although all people are born with flat feet, by age 12 or 13 most people's feet have formed metatarsal (side-to-side) and longitudinal (lengthwise) arches. These give spring to the step and help distribute body weight evenly across each foot and the lower limbs. Arches provide the feet with mobility and resiliency to adapt to different surfaces.

Some people inherit arches that never form and thus have lifelong joint problems and loss of balance. Stresses such as overweight, poor posture, and overuse of the feet may weaken the supporting ligaments and muscles, causing them to "fall." This condition can cause pain, inflammation, problems with the Achilles tendons, shin splints, stress fractures, and bunions.

To treat flat feet, the physician or podiatrist may prescribe an *orthotic*, a custom-designed arch support. The device redistributes the weight and aligns the foot for walking and absorbing excess shock.

Achilles Tendinitis

The Achilles tendon, located at the back of the heel, connects the foot to the muscles in the leg. The name comes from the Greek mythological warrior Achilles, whose mother wanted to protect her baby from being killed. She was told by the gods to dip him in the river Styx; no arrow would penetrate his body where the water had touched it. But she held him by the heels, and that area was not protected. Years later, he was killed when a poisoned arrow struck the back of his heel.

The Achilles tendon can be damaged from overuse, causing a painful injury. Conservative treatment, or R.I.C.E. (see Chapter 14) usually will cure Achilles tendinitis unless the cord is torn or ruptured. Surgery is required to correct the tear or rupture.

Toe Problems

The most common conditions affecting the toes are hammertoes, mallet toes, and overlapping toes. *Hammertoes* look like the claw of a hammer. The toe bends at the joint, and the tendon connecting the muscle to the bone and ligaments either tightens or stretches. Thus the bones assume an abnormal position that may be flexible or rigid. Often, poorly fitting shoes push on the top of the toe, causing pressure. If it is not treated, the condition may lead to ulcers and infection. *Mallet toes* are similar to hammertoes except that the deformity occurs near the end of the toe.

An *overlapping toe* is usually positioned at birth. The abnormality occurs in the soft tissue of the tendons, ligament, or skin. It often occurs on the fifth or little toe but may occur in any toe.

Metatarsalgia

The *metatarsals* are the five long bones in the foot. Each metatarsal has a narrow shaft and a knobby tip that links the anklebones, or tarsals, to the toe bones, or phalanges. *Metatarsalgia* literally means "pain in the metatarsals"; the term refers to a group of foot disorders. Symptoms include pain in the balls of the feet and a feeling of walking with rocks in the shoes. Although the condition can affect any age group from adolescents to older adults, most cases occur in middle-aged women.

The main cause seems to be the use of high-heeled shoes, which throw the body weight forward onto the balls of the feet. Many contributing factors may also have a role: rheumatoid arthritis, stress fractures, accumulation of fluid, fatigue, flat feet, and excess weight. Symptoms usually disappear with better-fitting shoes, but a physician or podiatrist may recommend anti-inflammatory medications or foot pads. Surgery is seldom necessary.

Bone Spurs

Bones spurs are sharp, pointed projections of bone that develop under the nail or at the heel. Occasionally, deformed nails are associated with a small spur at the tip of the toe under the nail, pressing on nerves and causing pain. The physician may make a small incision at the tip of the toe and insert a small instrument to reduce the spur.

The heel bone, or calcaneus, is the largest bone of the foot, projecting back beyond the leg to form a lever for the muscles of the calf. Bearing all the body's weight with each step, the heel receives the stress of any biomechanical fault such as flat feet, high arches, and tight Achilles tendon. Any upset in the normal mechanics causes tremendous stress on the *plantar fascia*, long bands of fibers that attach at the beginning of the toes and help create the arch. When a person gains weight, does high-impact athletic activities, or endures prolonged standing or walking, more stress is placed on the heel-end plantar fascia. A heel spur may form at the bottom of the calcaneus. Not all spurs hurt; inflammation of the structures causes the pain. Treatment includes rest; use of oral anti-inflammatory medication, heel cups, and orthotics; and surgical removal (as a last resort).

AMPUTATION

Throughout the history of orthopedics, the use of amputation (removal of a limb) to sever body parts afflicted with infection and disease has been a major occupation of doctors and surgeons. In long-ago wars, soldiers suffered from gangrene caused by the presence of the anaerobic bacteria clostridium perfringens, that required them to endure amputations without anesthesia and boiling oil being poured on the stump. Amputations today take place under very different conditions. Most amputations involve the extremities of the hands and feet, and possibly a limb.

The loss of a body part is potentially devastating, resulting in profound psychological, physical, and vocational consequences. In 1997 the *Southern Medical Journal* published a study of the kinds of amputations performed in U.S. hospitals from 1988 through 1996; the 1,199,111 amputations performed during that period were divided into five classes:

- *Trauma related.* Examples involve automobile accidents in which the limb is crushed, or gunshot wounds. Most of these involved the upper limbs—hands, arms, occasionally shoulders. (See photo.)

- *Congenital deficiency.* Some children are born with a deformity that must be removed.

- *Cancer related.* Conditions such as bone cancer sometimes require amputation.

- *Dysvascular.* Some conditions cut off blood circulation and cause gangrene. Diabetes is an example, usually requiring amputation below the knee.

- Of other origins.

When a person loses a limb, the possibility of using an artificial limb, or *prosthesis*, is considered. Although such a device can never substitute for the real arm or leg, the new prosthetics are very lifelike; some even work with the body's own electrical energy. The devices are fitted by

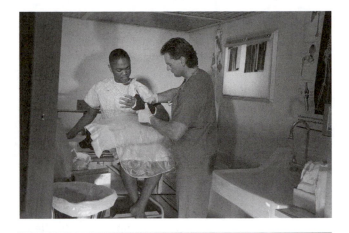

Doctor putting a cast on a boy's arm. © Photodisc Blue.

an orthotist, a specialist trained in making devices for the feet. Part of the recovery after surgery involves working with a physical therapist, who will help the patient strengthen the muscles in the other limb. Every case of amputation is individual, and the treatment and prosthetic are custom-made for that person.

FOCUS ON HANDS

In 1834, Sir Charles Bell (1774–1842), a Scottish anatomist, wrote a treatise called *The Hand* in which he expressed his awe at merely looking at the hand. One only has to observe the handiwork of an artist, a pianist, or a surgeon to appreciate what the hands can do. They are given flexibility by the wrist to accomplish precise action.

Once deformities are established, hands cannot be significantly altered by splinting, exercise, or other non-operative treatment. Deformities of the hand interfere with the person's psychological and occupational well-being, as well as the physical. Deformities of the hand include the following:

- *Mallet finger.* This is a flexion problem of the finger joint wherein the fingertip droops and cannot be extended. The problem arises from a tendon injury or destroyed bone at the fingertip.

- *Swan-neck deformity.* This is characteristic of rheumatoid arthritis, with several other causes such as the non-union of a broken finger. The person's fingers form a swan-neck shape that can interfere with pinching and grasping.

- *Erosive or inflammatory arthritis.* This is a clinical form of OA that causes bony enlargement of the joints in the hands known as Heberden's nodes and Bouchard's nodules.

Dupuytren's Contracture

This deformity was named after the celebrated French surgeon Baron Guillaume Dupuytren (1777–1835). In 1831, he studied this condition, which was prevalent in northern Europe, Scandinavia, and Russia. There is a strong tendency for the disorder to run in families, and some researchers propose that it is the result of a single gene mutation.

Although the exact cause of the condition is unknown, it is often accompanied by bursitis, arthritis, and tendinitis. The signs of Dupuytren's contracture are as follows (see Figure 16.2):

- *Nodules.* A lump (nodule) forms in the palm, most often at the base of the ring finger, little finger, or thumb. These usually are the first symptoms.

- *Pits.* A small but deep indentation appears in the skin of the palm or in a finger. The palmar fascia, the fibrous covering of the muscles on the palm, contract, and the skin tightens to form the pit.

- *Cords.* An unnatural cord of fiber extends from the palm into a finger, pulling the finger into a bent position. Eventually, all other fingers except

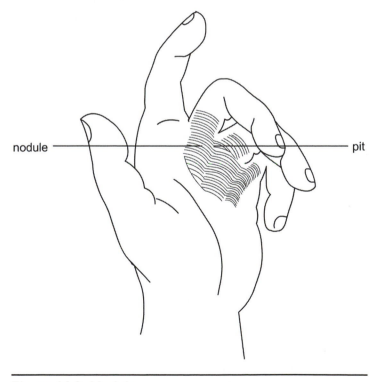

Figure 16.2. Nodule.

the index finger may be bent. One or two affected fingers may interfere with daily activities.

The process is slow. In 15 to 20 percent of the cases, the feet are also involved. The incidence is higher in men and increases after age 45. It is an inherited condition but is also associated with diabetes, alcoholism, and epilepsy.

Surgery may help restore maximum hand function but does not provide a cure. Following the surgery, the hand is immobilized in a large dressing like a boxing glove. A physical therapist will work with the patient to help regain proper function of the hand.

Ganglion Cyst

When a rubbery lump forms on the back of the hand or top of the foot, it is a *ganglion*. This results from an accumulation of a jelly-like substance that has leaked from a joint or tendon shaft. The bump, or *cyst*, forms under the skin at the wrist or on a finger, shoulder, elbow, or knee. Ganglion cysts usually increase in size but do not become cancerous. Generally they are painless, but they may be cumbersome and restrict movement.

What causes the cysts is not known. Occasionally they are associated with rheumatoid arthritis, but people in certain occupations that use the hand or wrist also have increased risk. A physician diagnoses the cyst with a physical examination as well as with x-rays. Draining the collected fluid is recommended.

FOCUS ON BACKS

The back, made up of thirty-three bones running from the skull to the coccyx, or tailbone, is a great feat of engineering. Vertebrae are stacked on top of each other like spools; in the spaces between are spongy disks, cushions that have a strong fibrous covering that protects a gel inside. A network of ligaments and muscles holds these structures together. The spine protects the delicate spinal cord of the central nervous system, which forms the nerve network of thirty-one spinal nerves extending to all parts of the body. Because of the complexity of the structure, the back can sustain a number of injuries and diseases. Even a simple change in balance or alignment—whether of disk, vertebra, or muscle—can cause painful problems.

Strains and Spasms

A simple ligament or muscle pull can cause pain at any point in the spinal column. However, the most common problem area is the lower back. Back discomforts can affect any age group. They usually subside after a reasonable period of rest.

Everyday occurrences may injure the back. It is important to develop simple prevention strategies to protect the back.

EXERCISE

"Bad backs" are generally caused by the side effects of inactivity. Out of shape, overweight persons put tremendous stress on their backs. It is especially important to develop strength in the muscles of the abdomen and buttocks because they support the lower back. The secret of exercise is to start slowly and make smart moves.

LIFTING

The legs, rather than the back, should do the work. Holding objects close to the body and avoiding lunging upward are also important.

SLEEP

Lying down can help rest an aching back, but the wrong position can aggravate pain. Three positions are recommended to prevent problems:

- Sleeping on the side with the legs drawn slightly forward, holding a pillow between the knees

- Sleeping on the stomach only if the stomach is cushioned with a pillow

- Sleeping on the back with the knees and neck supported with pillows

POSTURE

For proper posture, people should stand straight with an imaginary line running from ear to instep. Sitting erect is important. Because sitting may put more stress on the back than standing, it helps to use a straight chair or one that supports the small of the back and to take periodic breaks—even for a moment—from long stretches of sitting. Maintaining good posture at all times is crucial. Two common forms of poor posture are (1) a slouching position with the shoulders rolling forward, and (2) swayback.

Herniated Disk

A herniated disk is also called *prolapsed disk*, *ruptured disk*, *protruding disk*, and *slipped disk*. The latter term is wrongly named because there is no actual slippage. The condition occurs when a disk flattens out and presses on a nerve. Sometimes the person may have a worn facet where the rib joins the spine, and all these parts may press on a nerve, causing severe leg pain. Women in their late fifties are prone to disk problems, but less so into their sixties and beyond when the vertebrae stabilize.

If a disk ruptures, the gel in the spine may leak out, causing pressure on a spinal nerve. For most people, two to six weeks of bed rest—along with conservative treatment of heat or ice, massage, gentle exercise, or traction—may be enough for healing. One treatment involves an injection of chemopapain, an enzyme from the papaya tree. This injection causes the disk to shrink, reducing pressure on the nerve. Percutaneous diskectomy is a new procedure in which a large needle is inserted to remove the damaged disk. Conventional surgery is another option.

Whiplash

A car accident, a sports injury, or a sudden push from behind can cause the head to fly backward—an action called *hyperextension*—and then suddenly hurl forward—in *hyperflexion*. The snapping motion may upset cervical vertebrae in the neck, causing an array of symptoms that include stiffness, neck and shoulder pain, headache, and dizziness. The condition may not develop until quite a while after the occurrence. Professional help is necessary if the condition persists.

Neck Pain

The neck supports the head, a structure with the weight of a large bowling ball, and balances it on the top of the spine. Sometimes the neck becomes stiff or painful; this causes muscles to tighten and restrict all types of movement. Treatments for neck pain include the following:

- *Medication.* Nonsteroidal anti-inflammatory drugs are aspirin, ibuprofen, and naproxen.
- *Rest.* Complete inactivity may aggravate the pain, but lying down for brief periods may give relief.
- *Cold pack.* This can be applied for about 20 minutes each hour, but not directly on the skin. Ice packs should be wrapped in a cloth or towel.
- *Cervical collar.* This may help to restrict motion of the neck.

Conditions that do not improve after a few days require the care of a physician. People who have been in an automobile accident should have the condition checked, especially if they experience tingling or numbness.

Spondylosis

From the Greek *spondylos*, meaning "vertebrae," and *osis*, meaning "condition," *spondylosis* is sometimes called *degenerative joint disease* or *osteoarthritis of the spine*. The condition should not be confused with *ankylosing spondylitis*, an inflammation (see Chapter 16). Excessive use, injury, or the aging process may wear on the disks of the vertebrae and cause them to become narrow. Bone spurs may develop, resulting in stiffness and pain. Generally, the physician will recommend a combination of physical therapy, painkillers, and other self-help procedures.

Spinal Deformities

Three types of spinal deformities result in poor posture:

- *Abnormal lordosis.* The normal curve of front to back at the waistline is called *lordosis.* In abnormal lordosis the shoulder sits back too far, the rear end sticks out, and the person may have a condition known as swayback. Inadequate lordosis causes a flat back.

- *Abnormal kyphosis.* The normal sideways upper curve of the upper spine is called *kyphosis.* When it is abnormal, the person appears to be slouching. Sitting and standing in an abnormal slumping position can result in a permanent deformity.

- *Scheuerman's kyphosis.* Named for the doctor who first described this condition in 1920, this thoracic condition has two main types: a hereditary progressive type that is not painful, and a type located in the waist and lower spine that is painful.

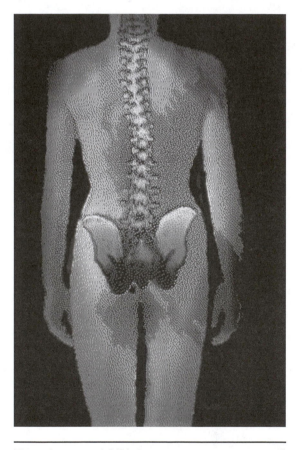

Three-dimensional MRI (magnetic resonance imaging) scan of scoliosis, a curvature of the spine in the dorso-lumbar region. Collection CNRI/Phototake.

Concentrating on maintaining good posture is important for good appearance and good health. Poor posture is a common cause of backache.

Scoliosis

Doctors throughout medical history have described scoliosis (see photo). Hippocrates (ca. 460–ca. 377 BCE) the Greek physician, named the condition based on the Greek word for "crooked." Scoliosis is a lateral, or sideways, curving of the spine that develops in late childhood or early adolescence (see Figure 16.3). The curves of scoliosis resemble the following patterns:

- A gentle curve like that of a parenthesis: (

- A curve that may cause the spine from behind to look like a lopsided S

- A right thoracic curve like a single C in the upper spine and bending to the right; this is the most common pattern in scoliosis.

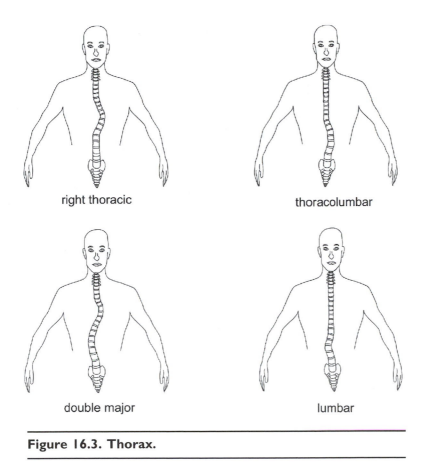

right thoracic   thoracolumbar

double major   lumbar

Figure 16.3. Thorax.

- A C-curve that starts in the chest and ends below the waist; called *thoracolumbar*
- A short C-curve in the lower back; called *lumbar*
- An S-curve where two parts are roughly equal

When the spine curves to one side, the vertebrae move around, or rotate, making ribs on the outside open up wide and ribs on the inside grow closer together. This causes a rib hump on one side of the back that is visible when the person bends over.

One cannot always tell by looking who has scoliosis. The curvature is slight at first and progresses as the child grows. Not all patients have a rib hump. The most visible signs are uneven shoulders and hips, revealed by slanting waistbands on skirts and pants.

About 80 to 90 percent of cases are known as *idiopathic*, meaning "having no known origin." Four types of scoliosis are idiopathic:

- *Infantile*: appears between birth and 3 years of age
- *Juvenile*: appears between age 3 and puberty
- *Adolescent*: the most common type
- *Adulthood*: a type that probably began earlier and was not spotted

About 15 percent of cases do have a known cause. *Congenital scoliosis* is a condition a person is born with: A trauma in the uterus causes a malformed vertebra called a *hemivertebra*—so called because it looks like a half-vertebra or a block.

For treating scoliosis, early detection is important. Legislation in most states requires screening of students in schools and referral to a physician, especially a family physician or pediatrician. If the doctor agrees with the screening assessment, he or she may refer the student to an orthopedist who specializes in scoliosis.

BRACES

In terms of straightening back and limb problems, people have always considered using an external device like a brace. Hippocrates first had the idea when he gave the condition a name. In the Middle Ages, the same tradesmen who equipped knights for battle made heavy metal corsets for people with scoliosis. The idea was that the brace would hold the spine rigid and prevent the scoliosis from progressing. Only the wealthy could afford these corsets, and they were not very successful in straightening the spine. But the idea of an external fixture persisted, and more progress has been made in the last fifty years than in all the preceding centuries.

A brace, or orthotic, is an appliance made of plastic or metal that supports the young spine to prevent the curve from progressing. Braces are recommended for children and teens, whose bones are still growing. Braces do not support the back; they force the spine into a straighter position.

Most braces have a pad held in place by metal rods or plastic plates. There are two kinds of braces:

- *Active, or dynamic, brace*—exerts pressure on the spine. The first of these braces was the Milwaukee brace, which has a neck ring that exerts a pull on the spine from the top. The combination of brace and exercise to develop the upper-body muscles help to hold the curve straight after the brace is removed.
- *Passive, or static, brace*—the model presently in favor. This type merely holds the spine in place and does not do a lot of the push-pull of the active brace.

Although wearing a brace is somewhat awkward, researchers claim it does work. One study conducted in Sweden followed the progress of 300 girls

Backpacks and Back Pain

Worldwide attention has focused on the role of backpacks in the development of low back pain in adolescents. Backpack loads in excess of 10 percent of body weight have been shown to increase stress on vertebra, causing the trunk to lean forward and resulting in decreased lung volume. Studies among French middle school and Australian high school students found an increase in complaints of back pain.

In the June 2003 issue of the journal *Spine*, a team of doctors published their findings from a study of 1,122 backpack users in American high and middle schools. Of the backpack users, 74.4 percent were classified as having back pain and were in poor health. Female gender and increased body mass were also associated with back pain. The study concluded that the use of backpacks was associated with back pain.

with adolescent idiopathic scoliosis. A control group of 131 participants did not wear braces, and 115 girls did. After five years, 70 percent of the girls who did not wear braces had progressed 6 degrees or more, in contrast to only 20 percent of the braced girls. (Correction of the spinal curve is measured in degrees.)

SURGERY

Surgery may be recommended for scoliosis under certain conditions. Surgery has been proven to correct the curvature. Most doctors agree that the magic number is 50 degrees. If the curve in the child's spine reaches this number, there is a strong likelihood it will continue to deteriorate after skeletal maturity has been achieved. Doctors formerly thought that the progression did not take place, but studies have shown that the deterioration does indeed continue.

If the curve is great or if back pain is severe, the person may consider surgery. Many surgeons are hesitant to operate on adults. Risks of surgery include the possibility of complications, but these are minimal—especially in children.

Awareness of care of the hands, feet, and back may make the difference in a life of pain and a life that is productive. Taking care to use common sense in daily activities like lifting and avoiding accidents can prevent potential problems of these important parts of the skeletal system.

17

Diseases and Disorders of Bones

Bones are living, changing tissues made up of proteins, sugars, minerals, and other materials. They produce blood cells and store the minerals calcium and phosphate. The complex activities that occur in the bones can lead to a wide range of diseases and disorders.

Some bone diseases are common, such as osteoporosis. Others, such as fibrous dysplasia and Paget's disease, are rare. Certain conditions, such as osteomalacia and rickets, are common in underdeveloped countries. Infections that cause inflammation disorders of the bones lead to conditions such as osteomyelitis. All these disorders result from improper functioning of some part of the skeletal system. People with these types of skeletal disorders are covered under the Americans with Disabilities Act (ADA). Children who have any type of disability are able to receive a "free and appropriate education" under the Individuals with Disabilities Education Act (IDEA) (see "Public Laws for People with Disabilities").

OSTEOPOROSIS

Osteoporosis is often called the silent killer because it has no symptoms. This common bone disease, which develops over a period of years, affects more than 25 million Americans, 80 percent—or 20 million—of whom are women. More than one in three women over age 50 will experience an osteoporotic fracture in her lifetime (see photo). The condition causes more than 1.3 million fractures each year, including 500,000 spinal fractures, 250 hip fractures, and 240,000 wrist fractures. Among postmenopausal women—those who have passed through menopause—osteoporosis is more prevalent than any other condition of older women.

Public Laws for People with Disabilities

At one time children with disabilities such as osteomalacia and genetic conditions such as brittle bone disease, spina bifida, and Cockayne syndrome were isolated and placed in separate institutions. A sweeping group of federal laws has changed this practice to require that all children now have an equal opportunity for the best free and appropriate education. The following laws have made a difference in the opportunities and quality of life for these children:

1973	The Vocational Act—Section 504 introduced rules for addressing needs of children with disabilities. States must provide all children a "free and appropriate education." Children with severe problems (such as those with tubes for bladder management) must be cared for during a regular school day and helped to attain their independence and potential.
1975	Educating All Handicapped Children Act (EAHCA)—This legislation became known as Public Law 94–142. It says that disabled children should be mainstreamed, or sent to a regular school, as much as possible. Implementation requires changes in school equipment and curriculum. Architectural changes, such as wheelchair-accessible elevators and ramps, are mandated.
1990	Americans with Disabilities Act (ADA)—This law prohibits any discrimination on the basis of disability.
1990	Individuals with Disabilities Education Act (IDEA)—As part of the ADA, this law ensures children's rights to an appropriate education. Each student has an individual education plan (IEP) developed by parents, teachers, and education specialists.
1997	The IDEA was amended to include more court cases and new regulations that strengthened the involvement of parents in developing the IEP.
2001	No Child Left Behind Act—Signed into law on January 8, 2002, this act requires sweeping reforms that include improved education and funding to support many initiatives, such as inclusion of children with disabilities in regular classrooms. Because of the influx of these students, teachers at all grade levels must receive special training to teach them.

Causes and Symptoms

Although many young women in their twenties hope they will never have the characteristic "dowager's hump" of a large number of older women (see Figure 17.1), they may be unknowingly sowing the seeds for the condition.

Constant dieting, inadequate intake of calcium, and over- or under-exercising begin the downward spiral. Called a lifestyle disease, osteoporosis may develop from a combination of some of the following risk factors:

- Early menopause (before age 40)

- Late menarche, or onset of menstruation at a late date (age 16 or over)

- Surgical removal of the ovaries

- Prolonged amenorrhea, or lack of menstruation—perhaps from rigorous athletic training

- Caucasian or Asian ancestry

- Family history of osteoporosis

- Small, slender stature

- History of fracturing easily

- Certain diseases such as rheumatoid arthritis, overactive thyroid, type I diabetes, liver or kidney disease

- Eating disorders such as anorexia nervosa and bulimia

- Diet low in calcium

- Sedentary life style

- Gastric surgery

- Heavy use of cigarettes or alcohol

- Prolonged use of steroids, thyroid hormones, anticonvulsants, or chemotherapy

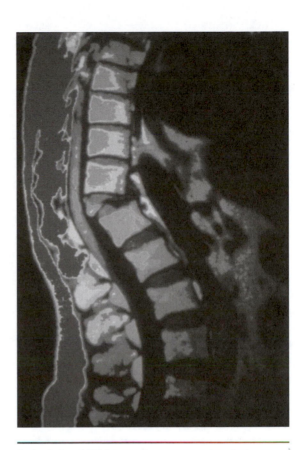

Sagittal-view MRI (magnetic resonance imaging) scan of a vertebral column with osteoporosis showing compression in the vertebra T12. Osteoporosis is characterized by a decrease in bone density producing porosity and fragility. Collection CNRI/Phototake.

Osteoporosis is a progressive disease of the skeleton caused by an imbalance in the body's bone-building cycle that results in loss of bone. The word comes from the Greek terms *osteon*, for "bone," and *porus*, for "pore." Osteoporosis literally makes the bones full of pores, or holes, and bone mass is reduced to the point that the bones become very fragile and thus prone to fracture. Bone mass is closely related to bone strength; the greater the bone mass, the stronger the bones and the less likely they are to break.

Scientists divide loss of bone mass into two conditions: osteopenia and osteoporosis. Although a certain amount of age-related loss of bone mass is

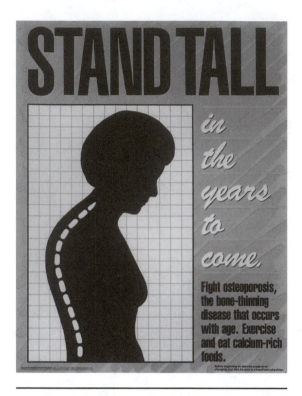

A poster warning women of the dangers of osteoporosis. © National Library of Medicine.

normal, the problem arises when the loss occurs at a faster pace than that of normal aging. Osteopenia is a condition in which the loss of bone mass begins to outpace the normal process. Just as high blood pressure is a risk factor for heart disease and stroke, osteopenia is a risk factor for osteoporosis. There are no symptoms.

The skeleton is made up of two kinds of bone: cortical and trabecular (see Chapter 5). Cortical bone looks solid and dense, with a circular pattern like that of cut wood; trabecular bone is surrounded by cortical bone, is porous, and looks like a honeycomb. Even though this type of bone appears delicate, it is in fact a very strong latticework made up of vertical and horizontal plates. Osteoporosis affects this microarchitecture. In the first five years after menopause, a woman can lose up to 25 percent of her total bone mass because of a drop in estrogen levels. Bone loss may then continue throughout life.

Actually, osteoporosis is a matter of degree. The bone does not differ from normal bone; there is just less of it. The following definitions express the relationship:

- *Normal bone:* Peak bone mass is 90 to 100 percent. Plates are thin and the microarchitecture is intact, making a low risk of fracture.

- *Osteopenia:* Peak bone mass is 75 to 90 percent. Plates are thinner and the microarchitecture is still intact, making a slightly higher risk of fracture.

- *Osteoporosis:* Peak bone mass is 75 percent or less. Plates are much thinner and the microarchitecture is disrupted, making a high risk of fracture.

Bones most affected are those of the spinal vertebrae, wrist, and hip. Visible effects occur in the spine. At age 40, there may be no signs of osteoporosis. At age 60, some small spinal fractures may occur—but still no symptoms. At age 70, the spinal fractures continue, and there can be a significant loss of height, a tilted rib cage, a protruding abdomen, and the development of a back hump called the *dowager's hump.* Loss of height

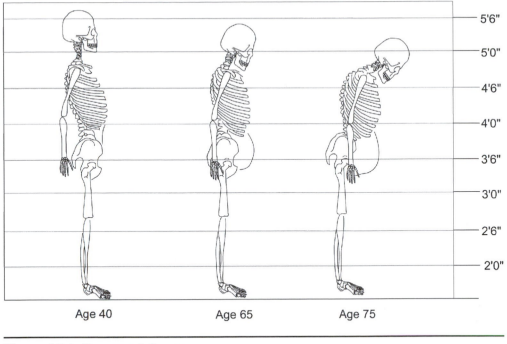

Age 40   Age 65   Age 75

Figure 17.1. Dowager's hump.

usually occurs in increments of one-half to one inch or more, depending on the number of vertebrae that have collapsed.

Men may also lose bone mass as they get older, but at a much slower rate than women. At skeletal maturity men have bigger bones and more mass than women, so they can usually withstand the age-related loss of bone mass.

Diagnosis of Bone Loss

At one time, little was known about the loss of bone mass until fractures occurred. There are now some accurate ways of determining whether an individual has lost bone mass. An instrument called a *densiometer* measures the relation of bone density, or the mineral content—mostly calcium—to the width of the bones. The densiometer sends radioactive x-rays into the bone and calculates how many rays are absorbed. The greater the absorption, the greater the bone mineral density, or BMD. Spine and hip bone are better predictors of fracture than wrist or heel. To predict an individual's fracture risk, technicians compare the reading to a chart of the average bone density in a population of normal young women at age 35, the time of skeletal maturity and peak bone mass.

Densiometers are used in two types of testing:

- *Single x-ray absorptiometry*, or SXA: A single beam of gamma rays measures bone density of the wrist or heel. This test is good because it is portable. A scanner passes over the bone and measures the amount of gamma rays that are absorbed. The procedure is painless.

- *Dual-energy x-ray absorptiometry*, or DEXA: This more accurate measurement is done in an office with a machine that passes over the spine and hip, recording bone density using two beams of gamma rays. The procedure is painless.

Prevention and Treatment

Strategies for both treatment and prevention are similar:

- *Diet:* Adequate amounts of calcium and vitamin D are necessary to build strong bones and a good supply of calcium reserves. Building good bone mass during the growing years (from birth to age 25) is an investment in the future. Calcium needs increase as one gets older. Menopausal women should take 1,500 mg of calcium and 400 International Units of vitamin D per day.

- *Exercise:* The specific effects of regular exercise on bones are not clear, but scientists do know that inactivity causes severe bone loss. Regular exercise is essential for osteoporosis prevention. The exercise should include both strength-building exercises and aerobic exercises, such as walking.

- *Medication:* Many physicians have recommended estrogen replacement therapy as a preventative measure. However, this practice is controversial because recent studies indicate that it might increase the risk of breast cancer. Several new medications, including a class of drugs called *bisphosphonates* that reduce bone resorption, are now approved by the Food and Drug Administration.

- *Surgery:* Surgery may be required in extreme cases, such as disk replacement.

PAGET'S DISEASE OF BONE

When the composer Ludwig van Beethoven (1770–1827) began to lose his hearing, he noted that some bones in his head began to become abnormally large. Determined to work and write music anyway, he composed some of his greatest works without the benefit of hearing. Beethoven might have been a victim of Paget's disease. Named for the nineteenth-century British physician Sir James Paget (1814–1899), the condition is also known as *osteitis deformans*.

Most people have never heard of Paget's disease even though it is the second most common bone disorder (after osteoporosis) among people over age 50. It may occur in up to 3 percent of Americans over age 60 and is rarely diagnosed in individuals younger than 40. In fact, many people who have the disease do not realize it. They mistake it for arthritis or some other condition. Often it is discovered accidentally when the individual has an x-ray for some other problem.

Causes and Symptoms

Paget's disease of bone is a chronic condition that may result in fragile, enlarged, and deformed bones in one or more regions of the skeleton. Throughout life, bone is involved in the process of growth and remodeling; but in this disease, the body makes more bone than it can absorb. The excess bone is abnormally large and deformed, and it fits together in a haphazard manner. Normal bone is tight and overlapping, like a well-constructed brick wall. In Paget's disease, the bone cells are haphazard, like bricks that were just dumped together.

The disease seems to run in families. It is more common in certain populations, primarily those of Anglo-Saxon descent, but almost never in Asian populations. Men and women are about equally affected. Some scientists think the disease may be the result of a slow-growing virus.

Paget's disease may appear in only one or two bones or even in just a portion of a bone. Mild cases have few symptoms, but the most common is bone pain. Symptoms include the following:

- If the skull is affected, individuals may experience headaches and hearing loss. Vision may be affected, and the head may become enlarged. Teeth may begin to fall out.

- When the spine is involved, low back pain is a frequent complaint. The spinal canal narrows in Paget's disease, causing a condition—known as spinal stenosis—that can lead to a loss of sensation and movement. Fractures in the vertebrae may lead to curvature of the spine.

- If the disease is centered in the weight-bearing limbs of the legs, the bones may fracture, take longer to heal, and begin to bow or deform. Damage to cartilage may lead to arthritis.

The disease is diagnosed with x-rays and a blood test that measures the level of alkaline phosphates, a substance that is overproduced in the bones of Paget's patients. Occasionally, Paget's disease is associated with the development of **sarcoma,** a type of malignant tumor.

Treatment

Although there is no cure, several treatments are effective, especially if they are applied early. Two drug families are commonly prescribed:

- *Bisphosphonates*, which reduce bone resorption. A calcium supplement is usually recommended along with these drugs.

- *Calcitonin*, which reduces bone resorption This drug is given through daily self-administered injections or a nasal spray.

As with osteoporosis, exercise and diet are very important in the treatment of Paget's disease. Exercise helps maintain skeletal health, prevent weight gain, and promote joint mobility. Surgery may be recommended to help fractures heal in a better position, to replace damaged joints, or to realign deformed bones and relieve knee pain.

OSTEOMALACIA AND RICKETS

Causes and Symptoms

The Greek root words *osteon*, for "bone," and *malacia* for "softening," describe the condition that occurs from a loss of calcium in the skeleton. The bones become flexible and are gradually molded by forces such as bearing the weight of the body on the legs. Symptoms include bone pain, muscle weakness, and fractures with minimal trauma. When osteomalacia occurs in children, it is called *rickets* (see photo).

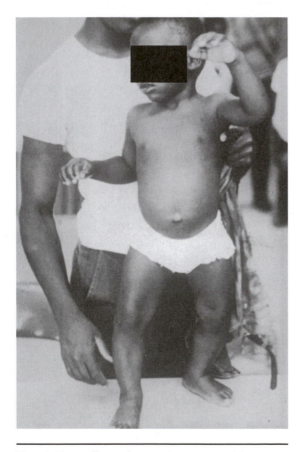

This child is suffering from malnutrition and has manifested symptoms of the disease nutritional rickets. This is a condition in which children's bones are too soft, and do not develop properly due to a deficiency of vitamin D. © The Centers for Disease Control.

Common causes of the condition include the following:

- *Improper absorption of fat*: This type of malabsorption, called *steatorrhea*, occurs when the body is unable to take in fats and thus passes them directly out in the stool. Because vitamin D is usually absorbed with fat, calcium is poorly absorbed also. Older adults may be at increased risk if they are lactose intolerant and inactive to the point that they avoid sunlight, which produces vitamin D in the body.

- *Increased amount of acid in bodily fluids*: When the kidneys are not functioning properly because of congenital or acquired kidney disorders, the increased acid causes the bones to gradually dissolve.

- *Inadequate processing of calcium*: For people who have trouble digesting milk products, long-term deficiency of calcium in the diet can cause osteomalacia as well as osteoporosis.

A hereditary form of rickets, called *vitamin-D resistant rickets,* is associated with shortened adult stature. Other conditions may also cause osteomalacia: kidney failure, kidney disease or cancer, and the side effects of medications that are used to treat epileptic seizures.

Diagnosis and Treatment

After blood tests and X-rays have determined the exact nature of the problem, oral supplements of vitamin D, calcium, and phosphorus are given, depending on the underlying cause. Improvement is usually evident within a week in people with a vitamin deficiency. In extreme cases, skeletal deformities may require surgical correction.

FIBROUS DYSPLASIA

Fibrous dysplasia is similar to Paget's disease in that abnormal growth of bone causes a haphazard arrangement of tissue cells. However, in this condition, instead of abnormal bone growth, the excess is in tissue fibers.

Causes and Symptoms

This chronic disorder of the skeleton causes expansion of one or more bones due to abnormal fibrous tissue that results in uneven growth, pain, brittleness, and deformity. Any bone may be affected. The term *monostotic* describes the effect on only one bone. The term *polystotic* describes the effect on numerous bones, usually on the same side of the body. The most commonly affected bones are the femur, tibia, ribs, skull, facial bones, humerus, pelvis, and vertebrae.

The condition usually occurs in children and young adults and is gender equal between boys and girls. The exact cause is not known, but it may involve a chemical abnormality in the protein of the bone. Symptoms include the following:

- A waddling walk
- Bone pain
- Bone deformity
- Bone fractures

In childhood, the condition may produce bone **lesions** (wounds or abnormal growths) changes in skin pigmentation, and endocrine abnormalities. The lesions sometimes stop at puberty.

In addition to monostotic and polystotic forms, there are three other forms of fibrous dysplasia:

- *McCune-Albright syndrome*: Named for two physicians who described it over fifty years ago, the condition occurs mostly in girls and is associated with abnormalities in bone disease with fracture, early puberty including menstrual bleeding, and splotches of skin pigment in an asymmetrical pattern. The skin patterns are called *café au lait*, or *coffee with milk*, because of their distinct color. The condition has a broad spectrum of severity.

- *Leontiasis ossea*: This form involves the skull and facial bones.

- *Cherubism*: A rare familial disorder, this type is characterized by enlargement of the upper and lower jawbones on both sides of the face. This painless, genetic disorder causes an overgrowth of fibrous tissue around the jawbones so that the child looks like a chubby-cheeked cherub. The disorder appears at age 3 or 4 and continues until the person reaches age 15 to 20, when the size of the jaw stabilizes, then becomes smaller, and ultimately is barely noticeable.

Diagnosis and Treatment

The physician may diagnose by using an x-ray and bone **biopsies**, cutting out a fragment of bone for study. Depending on severity, the condition may be treated with bisphosphonates, one of the drugs used in treating Paget's disease and osteoporosis.

OSTEOMYELITIS

Causes and Symptoms

Osteomyelitis is an acute or chronic bone infection, usually caused by bacteria but sometimes by fungi. The infection that causes the condition often starts in another part of the body and spreads to the bone via the blood. Affected bone may have been predisposed to infection because of recent trauma or surgery. Sometimes the infection occurs in vertebrae and in the bones of the feet in individuals with diabetes. In children, it usually affects the **metaphysis** of the tibia or femur as well as other growing bones that have a rich supply of blood. In adults, vertebrae and the pelvis are most commonly affected. As in other infections, pus is present; it collects within the bone and can result in an abscess. The abscess then deprives the bone of its blood supply. The condition may occur as a result of intravenous drug use.

Patients with acute osteomyelitis of the peripheral skeleton often are febrile (have a fever) and have weight loss and fatigue, localized warmth, swelling, redness, and tenderness. With vertebral osteomyelitis, the patient usually is not febrile but experiences tremendous localized pain and muscle spasms that do not respond to treatment.

Diagnosis and Treatment

A physical examination will reveal bone tenderness, swelling, and redness. A bone scan will indicate the infected bone, and blood cultures may help identify the causative organism. Antibiotics are selected for the specific organism. Surgery may be needed to remove the dead tissue.

BONE TUMORS AND CANCER

Causes and Symptoms

A bone tumor occurs when bone cells grow abnormally. Tumors may be **benign** (they do not spread) or **malignant** (they do spread, or become cancerous). Malignant cancers spread to other parts of the body. The causes of the abnormal growth of bone tumors is unknown: heredity, trauma, and radiation are possible causes.

Tumors that are benign can still cause many problems. The most common benign tumor type is called *osteochondroma*, from the Greek words *osteon*, meaning "bone," *chondro*, meaning "cartilage," and *oma*, meaning "tumor." As the name implies, this abnormal growth contains both bone and cartilage. These tumors most often occur between the ages of 10 and 20. They are monitored periodically by x-rays and usually regress with no treatment.

The incidence of bone cancer is highest in families with a history of any kind of cancer. Approximately five cases per million occur in children each year.

Malignant bone tumors can arise from two sources: primary bone lesions and metastatic cancers. Primary bone lesions are rare, accounting for less than 1 percent of all malignant tumors and are more common among young men. Malignant bone tumors include the following:

- *Osteosarcoma:* This is the most common primary bone malignancy in children, with the knee and proximal humerus being common sites. This type of tumor is difficult to diagnose and may be painless.

- *Ewing's sarcoma:* This type occurs in children in the flat and axial pelvic bones and in the diaphyses, or central shafts, of long bones such as the fibula and femur. The condition is associated with fever, weight loss, and local tenderness and redness.

- *Fibrosarcoma:* This type occurs in connective tissue.

- *Chondrosarcoma:* This type can be a **primary** or **secondary lesion** and is usually associated with dull, deep pain. There are five subtypes of this sarcoma.

Metastatic bone cancers arise in other areas and migrate to the bone. These are called *secondary lesions* because they originate in another place in the body—usually the breast, lung, prostate, kidney, or thyroid. Metastatic bone cancers usually affect adults.

The common symptoms of a bone tumor are as follows:

- Bone pain that may worsen at night
- A palpable mass, or swelling, at the tumor site
- Bone fracture from just a slight blow

Some tumors have no symptoms.

Diagnosis and Treatment

Bone tumors are diagnosed in several ways: x-ray, bone biopsy, bone scan, or a number of tests that include blood study for increased alkaline phosphatase, an enzyme that is present in the bloodstream when cancer occurs. Benign tumors may not require treatment but must be assessed periodically for progression or regression to see if surgical removal is necessary. Primary malignant tumors are rare and require treatment at specialized centers. After biopsy, a combination of **chemotherapy** (treatment with drugs) and surgery is necessary. **Radiation therapy** (treatment with x-rays) may be used before and after the surgery. The treatment of metastatic tumors varies according to the primary tissue or organ involved. Radiation and chemotherapy may be combined with hormone therapy.

Because bones are living, changing tissues, many diseases and disorders may occur. Researchers are working to treat and possibly cure the causes of debilitating illnesses. Also, Congress passed the Americans with Disabilities Act to give these individuals rights in society.

18

Genetic and Congenital Disorders

Bone disorders sometimes occur as a result of genetic anomalies. Whereas in earlier centuries people of unusual bone structure were not fully accepted in society, today they are recognized as being just like everyone else—except they happen to have a unique disability. Those of extremely short stature, for example, are no longer referred to as *midgets*; the preferred name of the condition is *dwarfism*, or *little people*. The condition of being extremely tall is referred to as *giantism*. Both are examples of unusual genetic conditions of the skeletal system.

A GENETICS PRIMER

A number of disorders are caused by defects in genes, or chromosomal abnormalities. Chromosomes are tiny thread-like structures in most of the cells of the body with the exception of red blood cells and some cells in bone marrow. Each human has twenty-three pairs of chromosomes, or forty-six in all. One chromosome pair comes from the father and one from the mother. One pair of chromosomes, the **sex chromosomes**, follows special rules. Men have an X and Y chromosome, and women have two X chromosomes. The other twenty-two pairs are called **autosomes**.

On the chromosomes, genes are lined up like beads on a string. There are 30,000 to 35,000 human gene pairs that determine traits for every aspect of our physical and biochemical systems. The complete set of genes is called the *human genome*. The Human Genome Project, completed in the year 2000, has mapped the location of many of the genes on chromosomes. Some disorders are caused by a change, or mutation, in only one gene; others are caused by interaction among many genes. Some genes are given names by scientists. Some of the names relate to a pathway or chemical the genes control or regulate, and other names are made-up words. For example, the gene *noggin* regulates the bone growth hormone.

Scientists today know the genes are made of proteins. Current research is trying to determine exactly which proteins make up the genes and how they work. This science is called *proteomics*.

In considering the genetics of disorders of the skeletal system, identifying the mode of inheritance is essential. For example, the form or type of dwarfism depends on whether its gene is on a sex chromosome—the X or Y—or on an autosome. If it is on an autosome, the condition is said to be either *autosomal dominant* or *recessive*. **Dominant** means that the trait will be expressed in the offspring. **Recessive** means that both traits must be inherited from the parents to appear in the offspring. If the dominant gene is present, the trait (how the person appears) will be seen in the offspring. If two recessive genes are inherited, that trait will appear in the offspring. The exception is the X-linked recessive condition. A man who carries a changed, or mutated (abnormal), gene on his one and only X chromosome will show the condition, whereas a woman will carry a second X chromosome that masks the disorder but will be the carrier to pass the gene to her sons.

Understanding the different kinds of inheritance can be quite difficult. To help couples understand a disorder when they are considering having a baby, they may be referred to a genetic counselor. *Genetic counseling* is a profession that involves giving people information and helping them make decisions about the risk of having children with genetic disorders. (See "Genetic Counseling: Theory and Practice.")

DWARFISM

Dwarfism is a condition in which a person is abnormally short (see photos). One form of dwarfism, caused by low levels of pituitary growth hormone, affects the whole body. This person will be small but normally proportioned. However, most dwarf characteristics are disproportionate, characterized by normal hands but short arms. (See "Facts about Dwarfism.")

Achondroplasia

Achondroplasia comes from the Greek words meaning "without cartilage." This genetic condition is the most common cause of short stature. The arms and legs are disproportionately short; the body size is normal; and the head is large. The shortness is especially noticeable in the upper arms and thighs. The person may also have a prominent forehead, a flat or depressed area at the base of the nose, a protruding jaw, and poor dental structure. The

Genetic Counseling: Theory and Practice

Genetic counseling is a health care service that provides medical information about genetic disorders and risks. Many disorders of the skeletal system have a genetic component. Counselors may give prospective parents several options if they find the fetus is afflicted with a genetic disorder. The couple can then make decisions about family planning and health care.

Less than fifty years ago, people did not know much about their genetic inheritance, or how genes are passed from one generation to the next. With the Human Genome Project (completed in the year 2000), scientists began to slowly unravel the secrets of genes. The information is overwhelming, and many people do not understand what it means. Genetic counseling can help to explain relevant parts of the genetic information.

Genetic counselors are trained medical professionals with a master's degree from an accredited program. Although the information about diseases and genes is very factual, these counselors add a personal touch that applies to each client's situation. Ideally this occurs before an at-risk pregnancy happens; usually it occurs after a suspected problem is identified.

The counselor gathers all the information about the condition and then discusses available tests and facilities. The couple gives information about their attitude toward risk, family goals, and ethical and religious standards. The counselor also serves as an advocate to help the couple navigate the complex medical system. During this process the counselor prepares a pedigree, a way of visualizing genetic disease in the family. Confidentiality is very important because insurance companies have been known to deny insurance to people in the pedigree. The counselor may order the following tests:

- X-rays

- Ultrasound

- Urine analysis

- General physical examination

- A chromosome study of the parents

- DNA analysis

- Amniocentesis of fluid around the fetus

- Chorionic villus sampling to analyze the chromosomes of the fetus

- Fetal cell sampling

Some of these tests can be very expensive, ranging from hundreds to thousands of dollars. Most of the one hundred or so gene analyses that are available test conditions where only one gene is involved.

The court dwarf Diego Rodriguez de Silva. Etching based on a painting by Velasquez. © National Library of Medicine.

hands may be short and sturdy, with loose ligaments that cause double jointedness. Signs of dwarfism are apparent at birth, and the diagnosis is usually made at that time. The average height of an adult with achondroplasia is about 4'0". Intelligence is normal. Indeed, many dwarfs have successful careers in a range of professions.

Achondroplasia is an autosomal dominant inherited condition, meaning that one individual carrying a changed gene has a 50 percent—or one in two—chance of passing on the gene to offspring. Autosomal dominance shows the following patterns:

1. It occurs in every generation.
2. Men and women have the condition with equal frequency and severity.
3. Each child of an affected parent has a 50 percent chance of being affected.
4. If both parents are affected, the child will probably have a more severe form of the disease and will probably die soon after birth.

Fantasy Island, starring Herve Villechaize and Ricardo Montalban. ABC-TV/The Kobol Collection.

The majority of people with achondroplasia are born to average-size parents, indicating that a mutation occurred in one parent's genes. These families have a reduced chance that a second child will be born with the condition. The cause of achondroplasia is suspected to involve the gene that encodes the fibroblast growth receptor 3 (FGFR3) on chromosome 4. The normal function of this gene is to slow down formation of bone by inhibiting the rapid development

---

Facts about Dwarfism

- An estimated one million dwarfs, approximately 4'10" and under, live in the United States today.

- One in every 10,000 children is born with dwarfism.

- There are over 300 different types of dwarfism, each a separate syndrome having special characteristics and problems.

- Most dwarfs are born to average-size parents with no history of dwarfism in their families.

- Normal intelligence and life span can be expected for most types of dwarfism.

- Most individuals with genetic dwarfism require surgery or other treatment to correct or prevent bone deformities. These deformities can cause severe pain, mobility problems, and progressive crippling.

- With good medical care dwarfs can lead rewarding, productive lives and participate in almost every occupation imaginable.

---

of chondrocytes, the cells that produce cartilage. Without this normal function, the abnormal gene increases the activity of the growth of cartilage and severely limits the growth of bone.

Hypochondroplasia and Pseudoachondroplasia

Hypochondroplasia (meaning "possessing little cartilage") is another form of short-limb dwarfism that is caused by an autosomal dominant mutation. It differs from achondroplasia in that the diagnosis is often made later in childhood, around 2 to 4 years of age. Many of the features of achondroplasia are similar to hypochondroplasia but not as obvious.

Pseudoachondroplasia (meaning "false achondroplasia") is a form in which the child may appear normal at birth but displays a delay in walking or a waddling gait. The growth rate begins to slow, and a long trunk appears along with shortening of limbs. Osteoarthritis can occur early in childhood; the most important complications are orthopedic, requiring correction of hip and knees. The responsible gene, COMP (cartilage oligomeric protein), is inherited in an autosomal dominant manner.

Diastrophic Dysplasia (DTD)

DTD, a rare growth disorder, was described in 1960; before then, it was thought to be achondroplasia with clubfoot. The word *dysplasia* signifies a deformed condition that is present at birth. But DTD is a distinct condition involving different genes, as well as some unique manifestations.

Because DTD is particularly present in Finnish populations, studies from that country have aided in locating the gene on chromosome 5, which encodes for sulfur—an important element for skeletal joints.

The condition is inherited as an autosomal recessive gene and has the following characteristics:

1. It appears in one generation and not in the parents or offspring.
2. Men and women are equally affected.
3. Both parents may be carriers with no symptoms.
4. Any offspring of an affected individual will be a carrier.

Orthopedic problems are common. Joints can be dislocated, especially at the shoulder, elbow, hips, and knees.

Deformities among the Amish

Because the Old Order Amish (a religious community founded in the seventeenth century that preserves a traditional way of living) is small and isolated in certain areas of the country such as Lancaster, Pennsylvania, it offers a rare opportunity to study the inheritance of certain disorders. Two conditions—cartilage hair hypodysplasia (CHH) and Ellis–van Crefeld (EvC) syndrome—are found in this population. Characterized by short limb stature, sparse hair, impaired immunity, and anemia, CHH individuals have a long trunk in relation to short limbs. Ellis–van Crefeld syndrome, another form of short-limbed dwarfism, includes additional characteristics. Diagnosed at birth, children with EvC have shortening in the middle parts of the limbs that shows most visibly in the lower limbs. As these children get older, they develop knock knees and may require surgery. Their hands are short and wide and have an additional digit next to the fifth finger, a condition known as *polydactyly*. The fingernails and toenails are small and have an unusual appearance; a harelip and partial cleft palate may also be present. Often, the teeth have erupted before birth. Found in 2000, the gene causing the syndrome has been mapped to the short arm of chromosome 4. It is inherited in an autosomal dominant manner.

Collagen Conditions

Collagen is a protein that is a major part of bone, cartilage, and connective tissue. Changes in type II collagen—a major component of the spine, cartilage, and vitreous humor (the fluid present in the eye)—may cause spinal changes resulting in a short-trunked form of dwarfism, as well as myopia (nearsightedness) and retinal degeneration.

X-Chromosome–Linked Conditions

Several rare types of dwarfism are carried on the X chromosome. The conditions may be dominant or recessive. Men are almost exclusively affected. The following disorders are included:

- *Hunter's syndrome*: This disorder affects the chemicals used in building connective tissues in the body. The condition is characterized by a large skull, coarse facial features, and mental retardation.

- *Chondroplasia punctata*: This condition involves a very rare form of shortness that can be inherited as X-linked dominant or recessive.

- *X-linked rickets*: In this type of rickets, calcium levels may be normal but there is a deficiency of phosphates in the blood owing to excretion of this element in the urine. Individuals have extremely bowed legs and short stature. The condition may be treated with phosphorus and activated vitamin D.

Missing Chromosome: Turner Syndrome

Turner syndrome involves a whole chromosome abnormality that affects about 1 in 2,500 girls. Girls with Turner syndrome have only one X chromosome instead of two. They are usually sterile and do not undergo normal pubertal changes unless they are treated with sex hormones. These girls are short, although treatment with growth and sex hormones can help increase their height. Orthopedic problems arise because of deformities associated with wedging of the carpal bones and bowing of the radius and ulna. Knee abnormalities occur in up to 60 percent of affected individuals. The chin is very small and sits farther back than its normal position. These girls display normal intelligence, but most seem to have difficulty in math and spatial exercises.

GIANTISM

Giantism is a condition of abnormally increased height that usually results from excessive cartilage and bone formation at the epiphyseal plates of long bones. The most common type of giantism, pituitary giantism, results from excess secretion of pituitary growth hormone. However, the large stature of some individuals can result from genetic factors rather than from abnormal levels of growth hormone.

Acromegaly

Acromegaly is a chronic metabolic disorder caused by the presence of too much growth hormone, which results in the enlargement of body tissues including the bones of the face, jaw, skull, and feet. The high level of growth hormone causes several orthopedic problems including pain in joints, knees, and ankles; over the small joints of the foot; at the shoulders, elbows, and wrist; and carpal tunnel syndrome. Acromegaly is usually caused by a benign tumor of the pituitary gland, which may have to be removed by microsurgery.

Marfan Syndrome

Marfan syndrome is a connective tissue disorder that affects the skeleton, lungs, eyes, heart, and blood vessels. Characterized by very long limbs, it is believed to have affected President Abraham Lincoln. Skeletal manifestations include curvature of the spine (scoliosis), abnormally shaped chest, loose jointedness, and disproportionate growth resulting in excessively tall stature.

In 1991, the condition was identified as an autosomal dominant disorder that is linked to the FBNI gene on chromosome 15. FBNI encodes for a protein called *fibrillin* that is essential for the formation of elastic fibers found in connective tissue. Without the structural support provided by fibrillin, these tissues are weakened.

Chromosomes and Giantism

Two other chromosomal aberrations contribute to giantism. An extra X in some women, called *Triple-X syndrome*, is responsible for very tall, thin stature. About 1 in every 1,000 women has an extra X chromosome. The person may experience menstrual irregularities and be less intelligent than her siblings, but rarely retarded. One man in every 1,000 has an extra Y chromosome. Ninety-six percent of these men appear normal, although they may experience acne as well as speech and reading problems. The XYY pattern is known as *Jacob syndrome*.

OTHER GENETIC DISORDERS

Osteogenesis Imperfecta (OI), or Brittle Bone Disease

A group of genetic disorders produce very brittle bones that are easily fractured. In this condition, type I collagen, which builds the scaffolding of the bones, is not properly formed. Fractures of the limbs before birth usually heal in poor alignment, causing limbs to be bent and short. Children with OI can have all types of fractures—spinal, rib, skull; incomplete and displaced. There are four types of OI, ranging from a mild form to one that is most severe and frequently lethal at or shortly after birth. The condition is caused by a dominant mutation. Recently, bisphosphonate drugs (the same type used for osteoporosis) have shown promise for treating OI.

Osteopetrosis

In this rare disorder the bones become overly dense, resulting from an imbalance between bone formation and bone breakdown. The condition has several names: *Alber-Schonberg disease*, *ivory bones*, and *marble bones*. There are several types of varying severity. Three forms exist—infantile, intermediate, and adult onset.

Fibrodysplasia Ossificans Progressive (FOP)

This extremely rare genetic disease causes muscle to be turned to bone. In the seventeenth century, the French physician Gui Patin (1601/2–1672) described a woman who had turned into wood. The wood that he described was actually new bone. Inherited through an autosomal dominant gene, FOP disrupts the mechanism that controls the growth of the skeleton. Any small injury to muscles, ligaments, and tendons may cause hard bone to form around the damaged area. Children are born with a malformation of the big toe and develop a "second skeleton" during childhood. Several genes are implicated. Treatment for this disease is to alleviate pain and symptoms; no cure exists.

Progeria

In this rare genetic condition, children undergo accelerated aging (see photo). The word *progeria* means "prematurely old"; the condition was described by Dr. Jonathan Hutchinson in 1886 and again in 1904 by Dr. Hastings Gilford. Thus it is sometimes referred to as *Hutchinson-Gilford progeria syndrome*, or HGPS. The total incidence is about one in four million children of all races and both sexes. About 100 cases have been identified worldwide. The affected children look like small old people and resemble each other more than their own family members. Characteristics are dwarfism, baldness, pinched

John Tacket, aged 15, with his parents in the background, talks about his illness, progeria, at a news conference in Washington, D.C., April 2003, to announce the discovery of the gene that causes this condition. © AP/Wide World Photos.

nose, delayed tooth formation, aged-looking skin, stiff joints, and hip dislocations.

On April 17, 2003, scientists announced they had discovered the gene responsible for progeria. The condition is caused by a mutation in the Lamin A gene found on chromosome 1.

Cockayne Syndrome

This rare disorder is characterized by sensitivity to sunlight, short stature, and premature aging. Edward Alfred Cocayne (1880–1956), a London physician who concentrated on hereditary diseases of children, found two classical forms of the condition. After exposure to sunlight, sufferers of the disease have a defect in the ability to repair deoxyribonucleic acid (DNA). Two genes have been implicated: CSA gene and CSB on chromosome 5.

Ehlers-Danlos Syndrome (EDS)

Ehlers-Danlos Syndrome is an inherited disorder that involves abnormal formation of the connective tissue. The cause of the disorder is an incorrect folding of collagen, the vital component in most tissues. Collagen is made of three strands wrapped into a triple helix according to a specific amino acid sequence. In EDS, a fault in the genes that program for collagen causes the strands to fold incorrectly.

People with EDS generally have a normal life span and normal intelligence but can throw their joints out of place, a condition called *double-joint-edness*. Depending on the severity, some people with EDS can throw several joints out of place and contort themselves into very unusual positions.

CONGENITAL MALFORMATIONS

The word *congenital* comes from two Latin words: *gen* meaning born and *con* meaning with. A congenital condition is one that the child has at birth. Some of these deformities might be caused by something that happens to the mother such as measles or conditions that cause high fever. Some are suspected to be of genetic origin, and others have no explanation.

Sometimes children are born with deformities or missing parts of limbs. The latter condition is usually more common in the upper limb than in the lower limb. Such conditions as *polydactyly*, or extra fingers; *syndactyly*, or webbed fingers; and *club hand*, where the radius bone on the thumb side is missing, may be seen. Similar disorders may occur in the feet.

Clubfoot

Clubfoot, a deformity that can be mild or severe, affects one or both feet. True clubfoot affects all the joints, tendons, and ligaments in the foot. In most cases it is *idiopathic*, or of unknown origin. If left untreated, the child will walk on the outside top surface of the foot and will ultimately experience difficulty in walking and other problems. The treatment to correct the deformity as early as possible involves the use of casts, physical therapy, or surgery.

Cleft Palate

A cleft palate is a congenital deformity in which the palatal plates (in the roof of the mouth) fail to close during the second month of fetal development, causing a hole in the roof of the mouth. The word *cleft* means "split" or "gap." Cleft palate and cleft lip are separate conditions that sometimes occur together. Approximately one in seven hundred to eight hundred births suffers from cleft lip, cleft palate, or both. Clefting is associated with a number of problems, including cosmetic deformities, dental problems, and difficulties in speech, swallowing, and growth.

Usually, the normal mouth develops in the following way:

- Two palates are present: the primary palate, the partition between the nasal cavity and mouth; the secondary palate, formed from the maxillary arches and frontal nasal process (see Chapter 6).

- Two halves of the palate meet in the middle. The floor of the mouth under the tongue grows together first, and by the eighth week the halves above the tongue form the palate. The dangling bit of tissue at the back of the mouth forms the uvula.

- The primary palate has no cleft at any time during fetal development, as the secondary palate does.

- From 6 to 9 weeks (some say 7 to 12 weeks), changes occur in the secondary palate that has not come together yet. The palatal shelves shift from a vertical, or growing down, position to a horizontal position. The tongue moves away from the shelves. The palatal shelves come together and fuse.

Cleft lip results from a failure of the epithelium, the layer of cells forming the skin, in the area of the primary palate. Clefts of the secondary palate result from the lack of fusion of the palatal shelves.

No one knows the reasons for the development of clefts. Some researchers think the tendency is inherited; however, clefts do not follow the rules of simple genetics and may involve many genes. The Native Americans of Montana have a high incidence of cleft palate, and similar types seem to run in families. Other researchers surmise that non-genetic factors such as disease, drugs, and alcohol affect the developing fetus and result in cleft. For example, a deficiency in tissue development at a certain stage during fetal growth may cause poor circulation to the face.

The treatment for cleft is usually planned soon after birth. A cleft lip is repaired by stitching the edges of the lips together, using flaps of skin from other

Reconstructive Surgery

Reconstructive surgery, or plastic surgery, is a branch of medicine dealing with the remodeling of damaged or deformed parts of the human body. This surgery is one of the oldest forms of known operations. Nose reconstruction was probably performed in ancient India as early as 2000 BCE. Susrata, an Indian surgeon, restored an amputated nose by using a flap of skin from the forehead.

Karl Ferdinand von Grafe (1787–1840), a German surgeon during the Napoleonic Wars (1800–1815), developed modern plastic surgery by improving upon previous ideas. An English surgeon, Joseph Carpue had adapted the ancient "Indian method" of using a flap of forehead skin. A sixteenth-century surgeon, Gasparo Tagliacozzi (1545–1599) had developed the "Italian method" of reconfiguring the nose by using a skin graft from the upper arm. Von Grafe combined these two approaches in his unique treatment. He developed an operation for repairing a cleft palate and also experimented with blood transfusion techniques that were necessary for his surgery.

Out of World War I came many advances in plastic surgery and new methods of treating severe wounds and facial injuries. Modern techniques of grafting tissue, skin, bone, and nerves reflect continual improvements in techniques and materials.

parts of the mouth. If the palate is cleft, the two halves are joined by using bone and skin grafts. Both conditions are repaired as early as possible to prevent speech problems from developing. Further surgery is carried out when the child is older to correct damage to the looks. Cleft lip is corrected when the baby reaches about 10 pounds at about 10 to 12 weeks; for cleft palate, the procedure begins between 6 months and 1 year of age. Nearly all clefts now are repaired in infancy, and generally by school age most children born with a cleft lip or palate are speaking well. (See "Reconstructive Surgery.")

Spina Bifida

Spina bifida is a birth defect in which a gap or split occurs when one or more vertebrae fail to form properly. The term comes from Latin, meaning "split spine" or "open spine." Spina bifida is one of a group of neural tube defects (NTDs) that occur in a portion of the backbone or vertebrae and surrounding tissues. The defects can affect any vertebrae but are most common at the base of the back, or lower spine. In the first three to four weeks after conception, the neural plate forms the neural tube, which contains the cells that make up the brain and spinal cord. In spina bifida, the sides of the plate fail to join properly at some point. The approximate timetable below follows the spinal defect as it develops in the embryo:

1. *About three weeks*: The neural plate fails to fold, and the sides do not join properly.

2. *Four weeks:* The embryo develops, but bone and muscle do not grow across the gap in the neural tube and cannot protect the nerves.

3. *Birth*: The spinal cord may protrude through the misformed vertebrae. Indications of the defect at birth are an abnormal tuft of hair, a collection of fat, and a thin membrane of skin covering the lower spine.

Spina bifida is the most common disabling birth defect. One in 1,000 children are born with a form of spina bifida. The condition may cause other problems in the legs, lower back, bladder, and bowels.

The cause of these neural tube defects is unknown. A tendency to run in families indicates there may be a genetic connection. However, 90 to 95 percent of families who have children with NTDs have no history of it. Mothers who have diabetes, who had a high fever during pregnancy, or who have taken certain epilepsy drugs to control seizures have a high incidence of children with NTDs.

In 1974, researchers found that pregnant women who carried fetuses with NTDs had a high level of the protein called *alpha-fetoprotein* (AFP) in the blood. This discovery led to the development of a blood test to detect the presence of NTDs before birth. Also, ultrasound examination may reveal the disorder. Couples whose tests reveal a high possibility of NTDs may be referred for genetic counseling (see "Genetic Counseling: Theory and Practice").

Treatment options depend on the type and severity. *Spina bifida occulta*, the least serious type, may not need any treatment. If problems occur as the child gets older, surgery may release the tethered spinal cord. The *meningocele* type may require surgery during infancy. The *myelomeningocele* type is a serious medical emergency requiring surgery within twenty-four hours of birth. The meningocele occurs when the meninges or covering of the spinal cord protrude through an opening in the spinal column; the myelomeningocele occurs when a portion of the spinal cord protrudes through an opening in the column.

In 1992, a study was published that indicated taking folic acid during pregnancy lowers the risk of these defects by 75 percent. Folic acid is a water-soluble B vitamin. During periods of rapid growth, such as pregnancy and fetal development when bones and other body parts are forming, the body requires more of this vitamin.

Congenital Dislocation of the Hip

Hippocrates (ca. 460–ca. 377 BCE) wrote in *De Articulis* of babies that suffer the greatest injury while in the uterus—displacement of the hip. In this condition, the baby is born with the ball-shaped head of the thighbone lying outside of its cup. The socket may be shallow and poorly formed, and one or both hips may be affected. The cause is unknown, but the disorder tends to run in families and occurs more in girls than in boys. About one baby out of sixty is born with the condition.

Throughout history, this condition has spurred doctors to develop all kinds of devices for correction. Splints are prescribed for six to nine months, and the child may have one or more operations to correct the condition. After the surgery, casts are put on the child's legs and must remain for several months. If it is treated early, the condition is almost always completely corrected.

The Human Genome Project gave great impetus for the study of genetic disorders. Scientists have pinpointed the gene or genes causing some disorders, but the majority of the causes of genetic and congenital disorders are still mysteries. However, researchers are convinced advances in molecular genetics will be the key for these conditions, as well as other disorders of the skeletal system.

Acronyms

ACL	anterior cruciate ligament	**COX-2**	cyclo-oxygenase-2 enzyme
ADA	Americans with Disabilities Act	**CTS**	carpal tunnel syndrome
AFP	alpha-fetoprotein	**CT**	computerized axial tomography
AI	adequate intake	**DEXA**	dual-energy x-ray absorptiometry
AIDS	acquired immunodeficiency syndrome	**DMARDs**	disease-modifying-anti-rheumatoid drugs
BMD	bone mineral density	**DNA**	deoxyribonucleic acid
BMP	bone morphogenetic protein	**DTD**	diastrophic dysplasia
BRDs	biologic response modifiers	**EAHCA**	Educating all Handicapped Children Act
CHH	cartilage hair hypodysplasia	**EDS**	Ehlers-Danlos syndrome
CNS	central nervous system	**ES**	embryonic stem cells
COMP	cartilage oligomeric protein	**EvC**	Ellis–van Crefeld syndrome

FDA	Food and Drug Administration	**NSAIDs**	nonsteroidal anti-inflammatory drugs
FGFR3	fibroblast growth receptor 3	**NTDs**	neural tube defects
FOP	fibrodysplasia ossificans progressive	**OA**	osteoarthritis
HGPS	Hutchinson-Gilford progeria syndrome	**OI**	osteogenesis imperfecta or brittle bone disease
HMSC	human mesenchymal stem cells	**OSD**	Osgood-Schlatter disease
IDEA	Individuals with Disabilities Education Act	**PCL**	posterior cruciate ligament
IEP	individual education plan	**PEMF**	pulsed electromagnetic fields
IU	International Units	**POP**	plaster of Paris
JA	juvenile arthritis	**PTH**	parathyroid hormone
LCPD	Legg-Calve-Perthes disease	**RA**	rheumatory arthritis
LD	lyme disease	**RDA**	required daily allowance
MAbs	monoclonal antibodies	**R.I.C.E.**	Rest, Ice, Compression, Elevation
MAGDC	Multiple Autoimmune Disease Genetics Consortium	**SCFE**	slipped capital femoral epiphysis
MCL	medial collateral ligament	**SLE**	systemic lupus erythematosis
MEMS	microelectrical systems	**SXA**	single x-ray absorptiometry
MRI	magnetic resonance imaging	**TB**	tuberculosis
NIH	National Institutes of Health	**TGF**	transforming growth factor
NMR	nuclear magnetic resonance	**TNF—alpha**	tumor necrosis factor—alpha

Glossary

Abduction Withdrawal of a part of the body from the body's axis

Accretion Increase by means of external addition or accumulation layer.

Adduction Movement of a limb toward the median line of the body

Adipocytes Cells that have large holes filled with fat

Amino acid A compound containing nitrogen that serves as the unit of structure of protein

Amphiarthrosis Joint that permits only slight movement

Anemia A condition that occurs when there is a shortage of oxygen-carrying red blood cells

Anterior Situated in front; at or toward the head end of a person or animal

Appendicular skeleton The skeletal structures composing and supporting the appendages; these include the bones of the shoulder and hip girdles as well as those of the arms and legs

Arthritis An inflammation of the joints.

Articulation A joint; the junction of two or more bones

Autograft A transplant of tissue from one part of a person's body to another

Autoimmunity Condition that occurs when the body defenses attack normal tissue

Autopsy An examination and dissection of a dead body

Autosomes Twenty-two non-sex-determining matched pairs of chromosomes that determine body characteristics

Axial skeleton The central supporting portion of the skeleton, composed of the skull, vertebral column, ribs, and breastbone

Barber-surgeons Men who cut hair and performed surgery, especially bloodletting and amputations

Benign tumor A growth that does not spread

Bilateral symmetry The property of having a body structure in which bisection through the vertebral column and the breastbone results in two halves that are mirror images of each other

Bioengineering Application of the principles of physics and engineering to the study of the human body

Biopsy Cutting out a fragment of tissue such as bone for study for possible cancer cells

Biotechnology The alteration of cells or biochemicals with a specific application, including monoclonal antibody technology, genetic engineering, transgenic technology, and gene targeting

Bipedal An animal that walks on two feet

Bone marrow The soft tissue contained in the central cavity of long bones and in the interstices of cancellous bones

Bonesetters Untrained wanderers who traveled around the countryside in Europe breaking bones and resetting them

Bursa A sac of fluid within a joint

Caduceus Staff of the Greek god Aesclepius that has one or two serpents coiled around it; used as a symbol of the medical profession

Callus A thickening of tissue resulting from the normal growth of cells

Cancellous bone Bone that has a latticework structure, such as the spongy tissue in the trabecular bone

Cancer A group of disorders resulting from the loss of cell control and the presence of abnormal growth

Cartilage A flexible supporting tissue composed of a nonliving matrix within which cells that have a nucleus are found

Cell A small complex amount of a living organism usually with a nucleus, cytoplasm, and cell membrane

Chromosome A structure within a cell's nucleus that carries genes; a continuous molecule of DNA with proteins wrapped around it

Collagen The albumin-like substance in connective tissue, cartilage, and bone

Compression forces Forces that squeeze items together; blows that press against the body

Condyle A rounded protuberance at the end of a bone where articulation occurs

Connective tissue A group of specialized cells that supports and holds together various parts of the body

Coronal plane Divides the body into front and back portions

Cortical bone The hard, dense bone that forms the outer shell of all bones

Costal Relating to the ribs

Cuboid bones Bones in the wrist that are shaped like cubes

Diaphysis The central shaft of a bone

Diathroses Joints allowing free movement

Differentiation The process during embryonic development whereby cells become specialized in structure and physiological function

Dislocation Condition when a bone is moved out of a joint

Distal Indicates direction away from the torso

DNA Deoxyribonucleic acid; the nucleic acid that bears coded genetic information

Dominant A trait or characteristic that is expressed in the offspring even though it is carried by one of the pair of genes

Dorsal Situated in back; the side along which the backbone passes

Encode Putting information in the form of a code

Enzyme An organic catalyst that causes a chemical change without being changed itself

Epiphyseal plate The cartilage mass between the epiphysis and the diaphysis

Epiphysis The portion of bone attached to another bone by a layer of cartilage

Equilibrium Balance

Erythrocytes Red blood cells

Estrogen A female sex hormone

Extension A stretching out, as in straightening a limb

Facets Pits or small, smooth surfaces on bone or other hard material

Fetal cell sorting A process whereby rare fetal cells are separated from a pregnant woman's blood

Fetus The unborn offspring of most mammals in the later stages of development

Flexion The bending of a joint or of body parts having joints

Fossa (s), Fossae (p) A hole or indentation

Gait analysis Scientific investigation of the way a person walks by developing a computer-generated profile using diodes or electronic sensors

Genes Units of heredity carried on chromosomes that determine a person's traits or characteristics

Genetic engineering Manipulation of genetic material; a broad term encompassing several biotechnologies

Genome All the genes present in an organism

Growth factor A protein that stimulates cell growth

Hematopoeisis The formation and maturation of blood cells

Hemoglobin Substance that gives blood its red color that contains iron and carries oxygen from the lungs to the body

Homeostasis The maintenance of a specific environment by the coordinated activities of various organ systems

Hormone A chemical messenger produced by an endocrine gland

Hydroxyapetite Mineral in living bone that forms a crystal structure

Inferior Direction given to a body part that indicates away from the head

Intracapsular ligaments Ligaments within the capsule at the joint

Ischemia Lack of blood supply to a tissue or organ

Joint The union between two bones

Lacunae Small depressions or spaces

Lamella (s), Lamellae (p) Thin scales or plates

Lateral Situated on a side

Lesions Wounds or abnormal growths

Ligament A tough band of connective tissue that connects bones to each other

Lymph A colorless fluid resembling blood plasma found in intercellular spaces and in small compact structures called nodes

Malignant tumor A cancerous growth that may spread throughout the body

Marrow *See* Bone marrow

Medial Toward the midline of the body

Membrane A thin, soft pliable layer of tissue covering an organ or structure

Meniscus (s), Menisci (p) Fibrous cartilage within a joint

Metaphysis Portion of a developing long bone between the diaphysis or shaft and epiphysis

Midsagittal plane An imaginary line that passes through the skull and spinal cord, dividing the body into equal halves

Monoclonal antibodies Single antibody types produced from a B cell, a type of white blood cell, fused to a cancer cell

Morphology Branch of biology that studies form and structure of the body

Mutation A change in a gene's biochemical makeup; a change in DNA

Necrosis Death of the cells

Notochord Column of the mesodermal layer in the embryo that becomes the axial skeleton of vertebrates

Opposable thumb In primates including humans the ability to use the thumb to touch each finger

Organ A part of the body composed of specialized tissues that perform specific functions

Osteoblast A cell involved in the formation of bone

Osteoclast A cell that destroys bone

Osteocytes Bone cells

Osteology The study of bones, from the Greek word *osteon*, meaning "bone" and the suffix *-ology*, meaning "study of."

Ovum A female reproductive cell; an egg

Palmar Indicates the palms of the hands

Parathyroid glands Four small, oval glands located on the thyroid gland that secrete a hormone for the control of calcium-phosphorus balance in the body

Periosteum The tissue that covers the outside of a bone

Pituitary gland An endocrine gland at the base of the brain that sends out growth hormones

Plantar Indicates the soles of the feet

Primary lesions Growths that begin at a particular site in the body; the initial place of origin of a growth

Primitive streak A band along the back of a three-week-old human embryo that forms an axis that other structures develop around and that eventually gives rise to the nervous system

Processes Projections or outgrowth of bone tissue

Protein One of a group of complex chemical compounds composed of carbon, hydrogen, oxygen, nitrogen, and other elements and characteristic of living matter

Proximal Indicates direction closer to the torso

Recessive A trait or characteristic that is expressed in offspring only when both pairs of genes are present

Resorption The process of absorbing again: describes the breaking down of bones

Rotation Involves turning a body part on an axis

Saggital plane An imaginary vertical line that divides the body into right and left segments

Sarcoma A tumor, usually malignant, in the connective tissue

Secondary lesions Cancers that have developed in one place and then spread to another area

Sex chromosomes Chromosomes that determine the sex of the child; men have X and Y chromosomes; women have XX chromosomes

Stem cells Primary germ layers in the embryo that develop into cells that will become parts of the body

Superior Direction given to a body part that indicates toward the head

Sutures Where the bones of the skull come together along serrated and interlocking lines

Symphysis A disk of cartilage where two bones meet

Synarthroses Non-moveable joints

Synovial fluid The clear fluid that is normally present in joint cavities

Tendon A tough band of connective fiber that attaches a muscle to a bone

Teratogenic Giving rise to a birth defect

Tissue A group of similar cells performing a specific function

Trabeculae Beams that act as strengthening girders of cancellous bone

Trabecular bone The porous, spongy bone that lines the bone marrow cavity and is surrounded by cortical bone

Transverse plane An imaginary line passing at right angles to both the front and midsection; a cross section

Vertebrate Animal with a backbone or vertebrae

Zygote A fertilized egg cell

Organizations and Web Sites

MEDICAL ORGANIZATIONS

American Academy of Orthopaedic Surgeons
6300 N. River Road
Rosemont, IL 60018-4262
Phone: 847-823-7186 or 800-346-AAOS
Fax: 847-823-8125
Email: pemr@aaos.org

Organization of professional orthopedic surgeons; a good source of information for general public.

American Lyme Disease Foundation, Inc.
Mill Pond Offices
293 Route 100
Somers, NY 10589
Phone: 914-277-6970
Fax: 914-277-6974
Email: inquire@aldf.com

Excellent site with maps and information on Lyme Disease.

American Osteopathic Association
142 E. Ontario Street
Chicago, IL 60611
Phone: 312-202-8000 or 800-621-1773
Fax: 312-202-8200
Email: info@aoa-net.org
www.aoa-net.org

General information on osteopathy.

Arthritis Foundation
P.O. Box 7669
Atlanta, GA 30357-0669
Phone: 800-283-7800
www.arthritis.org

Excellent source for general public; many publications available.

Cleft Palate Foundation
104 S. Estes Drive, Suite 204
Chapel Hill, NC 27514
Phone: 919-933-9044
Email: info@cleftline.org
www.cleftline.org

Good information with photos of surgery.

Ehlers-Danlos National Foundation
6399 Wiltshire Boulevard, Suite 200
Los Angeles, CA 90048
Phone: 323-651-3038
Fax: 323-651-1366
Email: staff@ednf.org
www.ednf.org

Good site for those interested in this rare condition.

Foresight Institute
P.O. Box 61058
Palo Alto, CA 94306
Phone: 650-917-1122
Fax: 650-917-1123
Email: foresight@foresight.org
www.foresight.org

Outstanding site for information on nanotechnology and nanomedicine; many links.

Little People of America, Inc.
P.O. Box 65030
Lubbock, TX 79464-5030
Phone: 806-687-1840 or 888-LPA-2001
Fax: 806-687-6237
Email: LPADataBase@juno.com
www.lpaonline.org

Information on growth-related problems; many brochures.

March of Dimes
1275 Mamaroneck Avenue
White Plains, NY 10605
www.marchofdimes.com

Excellent source for genetic disorders; very comprehensive.

Scleroderma Foundation
12 Kent Way, Suite 101
Byfield, MA 01922
Phone: 978-463-5843 or 800-722-HOPE (4673)
Email: sfinfo@scleroderma.org
www.scleroderma.org

Good source for information on scleroderma.

Scoliosis Association
P.O. Box 811705
Boca Raton, FL 33481-1705
Phone: 561-994-4435 or 800-800-0669
Fax: 561-994-2455
Email: normlipin@aol.com
www.scoliosis-assoc.org

Provides information and support to patients and families.

Scoliosis Research Society
611 E. Wells Street
Milwaukee, WI 53202-3892
Phone: 414-289-9107
Fax: 414-276-3349
Email: tjackson@execinc.com
www.srs.org

Dedicated to research into the causes of scoliosis.

Spina Bifida Association of America
4590 MacArthur Boulevard, NW, Suite 250
Washington, DC 20007-4226
Phone: 202-944-3285 or 800-621-3141
Fax: 202-944-3295
Email: sbaa@sbaa.org

Information about neural tuve defects, research, and publications.

WEB SITES

Askphysician.com
www.askphysician.com

This general orthopedic site contains lots of information on diseases and disorders.

Biomechanics Research Laboratory, The Johns Hopkins University
www.biomech.jhu.edu

A fascinating site that gives the biomechanics of baseball mechanics and knee kinematics.

Bonetumor.org: The Web's Most Comprehensive Source for Bone Tumor Information
http://bonetumor.org

A comprehensive resource on bone tumors of all kinds.

FreeOrtho.com
http://freeortho.com
This excellent site provides general information on orthopedics.

March of Dimes
www.mod.org
Provides information about certain problems of the skeletal system.

Medscape, Infocus: Orthopaedics
medscape.com/Home/Topics/orthopedics/orthopedics.html
This subsection of medscape.com gives current information on research articles on bones and orthopedics.

Merck
www.merck.org
The writers of the classic *Merck Manual* offer advanced information on diseases and disorders by systems.

Orthoguide.com
http://orthoguide.com/ortho
This orthopedic search engine has links for biomaterials divided into specialities.

Orthoworld: The Orthopaedics Internet Portal
www.orthoworld.com
This site is an excellent starting point for Internet-based orthopedics research. Some parts require membership.

United States National Library of Medicine
www.nlm.nih.gov
For conditions, click on Medlineplus.

University of Washington Orthopaedics & Sports Medicine
www.orthop.washington.edu
University of Washington Bone and Joint Sources have great general information with sound bites and movie clips.

What You Need to Know about Orthopaedics
http://orthopedics.about.com
This general orthopedic site has abundant information on diseases and disorders.

WideSmiles: Cleft Lip and Palate Resource
www.widesmiles.org
Hundreds of files on cleft lip and cleft palate.

WorldOrtho, the World of Orthopaedics, Trauma, and Sports Medicine
www.worldortho.com
This site for medical professionals in various fields has an online textbook, pictures, and other tools and a section on history.

Bibliography

"Advances: 1987–1997. A Decade of Research." *Arthritis Today* (1998).

Applegate, Edith J. *The Anatomy and Physiology Learning System: Textbook*, 2nd ed. Philadelphia: W.B. Saunders, 2000.

Arthritis Fact Book for the Media. Atlanta: Arthritis Foundation, 1986.

Arthritis Foundation Media Kit. Atlanta: Arthritis Foundation, 1994.

Asimov, Isaac. *The Human Body: Its Structure and Operation*, rev. ed. New York: Mentor, 1992.

"The Bone Morphogenetic Proteins (BMPs)." National Institute of Dental and Craniofacial Research. http://nidr.nih.gov/spectrum/NIDR4/4textsec2.htm.

"Bone Tumors." Medical Consumer Guide Web Site. http://www.medical consumerguide.com/primary_care/arthritis_musculoskeletal_disorders/bone_tumors.html.

Brooks, David, update. "Fibrous Dysplasia." December 3, 2001. Medline Plus at the National Library of Medicine. http://nlm.nih.gov/medlineplus/ency/article/001234.htm.

Charkin, H. *Children with Facial Difference: A Parent's Guide*. Bethesda, MD: Woodbine House, 1996.

Chu, Sharon. "Histology of Bone." Electronic Textbook at WorldOrtho. http://www.worldortho.com/database/etext/basicsci.html.

Cornett, Frederick D., and Pauline Gratz. *Modern Human Physiology*. New York: Holt, Rinehart, and Winston, 1982.

"Deformities." Merck & Co., Inc. http://merck.com/pubs/mmanual/section5/chapter61/61a.htm.

DeGroot Henry III. "On-Line Atlas of Bone Tumors." Bonetumor.org. http://bonetumor.org/tumors/pages/page7.html.

Dillingham, Timothy. R., Liliana E. Pezzin, and Ellen J. MacKenzie. "Limb Amputation and Limb Deficiency: Epidemiology and Recent Trends in the United States." *Southern Medical Journal* 95, no. 8 (2002): 833–875.

Dowling, Mike. "Don Johanson and Lucy." Prehistory at Mr. Dowling's Electronic Passport. http://www.mrdowling.com/602-lucy.html.

"Dupuytren's." The Indiana Hand Center. http://www.indianahandcenter.com /med_dup.html.

Echols, Michael. "The Civil War Federal Army Surgeon." Personal research notes. http://www.braceface.com/medical/Pages/Civil%20War%20Army%20Surgeon. htm.

"Ehlers-Danlos Syndrome." Disease Directory. http://diseasedir.org.uk/genetic/ gene0101.htm.

"Ehlers-Danlos Syndrome." Merck Source. http://www.mercksource.com /pp/us/cns/ cns_hl_adam.jspzQzpgzEzzSzppdoczSzuszSzuszS.

Eisenpreis, Bettijane. *Coping with Scoliosis.* New York: Rosen, 1998.

"Fibrous Dysplasia." Bone Diseases at the University of Maryland Medical School. http://www.umm.edu/bone/fibrdys.htm.

"Fibrous Dysplasia Support Online." Fibrous Dysplasia Online. http://fdsol.org.

Frank, Paul. J. "Scleroderma." *Dermatology Online Journal*, 7, no. 1 (2001): 16. http://dermatology.cdlib.org/.

"Functions of the Skeletal System." Project Skeletal. http://members.tripod.com/ projectskeletal/Functions.htm.

Gibbs, John. "Man as Machine." *University of Florida Today* 2, no. 3 (September 1986): 4–7.

Goldberg, Kathy E., and the Editors of U.S. News Books. *The Skeleton: Fantastic Framework.* Washington, DC: U.S. News Books, 1982.

Goodsell, David S. "Collagen." Research Collaboratory for Structural Bio informatics Protein Data Bank. http://www.rcsb.org/pdb/molecules/pdb4._1.html.

Gray, Henry. *Anatomy of the Human Body*, 20th ed. Edited by Warren H. Lewis. New York: Bartleby.com, 2000. http://www.bartleby.com/107/1.html.

Head, Susan, Bethany Thomas, and Gena Senibaldi. "Thalidomide Side Effects." Anatomy & Physiology at Richland College. http://www.rlc.dcccd.edu/ MATHSCI/reynolds/thalidomide/effects/effects.htm.

Hecht, Annabel. "Hocus-Pocus as Applied to Arthritis." *FDA Consumer* (September 1980).

Hill, Mark. "Development of the Musculoskeletal System." Embryology Program at the School of Medical Sciences, University of New South Wales. http:// anatomy.med.unsw.edu.au/CBL/Embryo/Notes/skmus.htm.

"History of Orthopaedics Menu." WorldOrtho. http://www.worldortho.com/pg2. html.

Howard, Scott, updater. "Bone Tumors." July 17, 2002. National Library of Medicine Medical Encyclopedia. http://www.nlm.nih.gov/medlineplus/ency/article/ 001230.htm.

"Human Osteology." Knowledge Weavers at the University of Utah Health Sciences Center. http://medlib.med.utah.edu/kw/osteo/osteology/bone types.html.

Hutchins, Amy, Kanitra Heynes, and Kim Creecy. "The Future of Thalidomide." Anatomy and Physiology at Richland College. http://www.rlc.dcccd.edu/MATH SCI/reynolds/thalidomide/future/future.htm.

"Johanson, Donald C. Biography." Ann Devlin Online. http://annonline.com/inter views/961294/biography.html.

Johnson, D.R. "Introductory Anatomy: Bones." University of Leeds School of Biomedical Sciences. http://leeds.ac.uk.chb/lectures/anatomy/3.html.

Johnson, D.R. "Introductory Anatomy: Joints." University of Leeds School of Biomedical Sciences. http://www.leeds.ac.uk/chb/lectures/anatomy4.html.

"Joints." Project Skeletal. http://members.tripod.com/projectskeletal/joints.htm.

"Legg-Calve-Perthes Disease." American Academy of Orthopaedic Surgeons. http://orthoinfo.aaos.org.fact/thr_report.cfm?Thread_ID=159&topcategory =About%20Orthopaedics.

"(Legg-Calve) Perthes' Disease." Orthoseek. http://www.orthoseek.com/articles/ perthes.html.

Lerner, K. Lee, and Brenda Wilmoth Lerner, eds. *World of Genetics*. Vol. 1. Detroit: Gale Group, 2002.

Le Vey, David. *The History of Orthopaedics: An Account of the Study and Practice of Orthopaedics from the Earliest Times to the Modern Era*. Park Ridge, NJ: Parthenon, 1990.

Lewis, Ricki. *Human Genetics: Concepts and Applications*. Dubuque, IA: Wm. C. Brown, 1997.

Lingham, Alex. "Some Effects of Thalidomide." Bristol University School of Chemistry. http://www.chm.bris.ac.uk/motm/thalidomide/effects.html.

Lipman, Karen S. *Don't Despair Cleft Repair*. Naples, FL: Lipman Productions, 1996.

"Lower Extremities." Human Anatomy at the Minnesota State University at Mankato. http://www.anthro.mankato.msus.edu/biology/humananatomy/skeletal/leg/ leg.html.

"Lyme Disease." American Lyme Disease Foundation. http://www.aldf.com/Lyme.asp.

"Lyme Disease Home Page." Division of Vector-Borne Infectious Diseases at the Centers for Disease Control and Prevention. http://www.cdc.gov/ncidod/dvbid/ lyme/.

Magner, Lois N. *A History of Medicine*. New York: Marcel Dekker, 1992.

Maples, William R., and Michael Browning. *Dead Men Do Tell Tales*. New York: Doubleday, 1994.

Meisterling, Robert, Eric J. Wall, and Michael R. Meisterling. "Coping with Osgood-Schlatter Disease." *The Physician and Sportsmedicine* 26, no. 3. (1998) http://www.physsportsmed.com/issues/1998/03mar/wall_pa.htm.

Notelovitz, Morris, and Marsha Ware. *Stand Tall! Every Woman's Guide to Preventing Osteoporosis*. Toronto: Bantam Books, 1985.

Notelovitz, Morris, with Marsha Ware and Diana Tonnessen. *Stand Tall! Every Woman's Guide to Preventing and Treating Osteoporosis*. 2nd ed. Gainesville, FL: Triad, 1998.

Oliveaux, Julie, Curator. "Dynamics of Calcium Metabolism and Bone Tissue." National Aeronautics and Space Administration (NASA) Human Space Flight web page. http://spaceflight.nasa.gov/history/shuttle_mir/science/hls/music/ scdynam.htm.

"Osgood-Schlatter Disease: A Cause of Knee Pain in Children." American Academy of Family Physicians. http://familydoctor.org/handouts/135.html.

"Osgood-Schlatter Disease (Knee Pain)." American Academy of Orthopaedic Surgeons. http://orthoinfo.aaos.org/fact/thr_report.cfm?Thread_ID=145&topcategory =Knee.

Peason, Helen. "True Stem Cell Found?" May 4, 2001. Nature Science Update at the Nature Publishing Group. http://www.nature.com/nsu/010510/010510-2.html.

Porter, Roy. *The Greatest Benefit to Mankind: A Medical History of Humanity*. New York: W. W. Norton, 1998.

Porter, Roy, ed. *Medicine: A History of Healing: Ancient Traditions to Modern Practices*. London: Michael O'Mara Books, 1997.

Russo, Eugene. "A New Approach to Autoimmune Disease." *The Scientist*, 17, no. 9 (May 5, 2003).

Sandler, Adrian. *Living with Spina Bifida: A Guide for Families and Professionals.* Chapel Hill: University of North Carolina Press, 1997.

Schlager, Neil, ed. *Science and Its Time: Understanding the Social Significance of Scientific Discovery.* 8 vols. Detroit: Gale Group, 2000–2001.

Sheir-Neiss, Geraldine, et al. "The Association of Backpack Use and Back Pain in Adolescents." *Spine* 28, no. 9 (May 1, 2003): 922–940.

Sherry, Eugene. "Bone and Soft Tissue Tumours." Lecture for Western Clinical School of the University of Sydney, Australia. WorldOrtho. http://www.world ortho.com/database/lectures/lecture5.html.

Shiel, William C., Jr. "Costochondritis & Tietze Syndrome." MedicineNet. http://www.medicinenet.com/Costochondritis_and_Tietze=_Syndrome/article.htm.

Siegel, Irwin M. *All About Bone: An Owner's Manual.* New York: Demos, 1998.

"Skeletal Pioneers." Project Skeletal. http://members.tripod.com/projectskeletal/Scientists.htm.

"Slipped Capital Femoral Epiphysis." American Academy of Family Physicians. http://familydoctor.org/handouts/282.html

"Slipped Capital Femoral Epiphysis." American Academy of Orthopaedic Surgeons. http://orthoinfo.aaos.org/fact/thr_report.cfm?Thread_ID=160&topcategory=Hip.

"Sprains and Strains." *Mayo Clinic Newsletter* 9, no. 9 (September 1991):1.

Steele, D. Gentry, and Claude A. Bramblett. *The Anatomy and Biology of the Human Skeleton.* College Station: Texas A&M University Press, 1988.

Tank, Patrick W. "All Bones—Organized by Region." Gross Anatomy at the University of Arkansas for Medical Sciences. http://anatomy.uams.edu/HTML.pages/anatomyhtml/bones_alpha.html.

———. "All Joints and Ligaments—Organized by Region." Gross Anatomy at the University of Arkansas for Medical Sciences. http://anatomy.uams.edu/HTML.pages/anatomyhtml/joints_alpha.html.

Temkin, Owsei, and C. Lilian Temkin. *Ancient Medicine; Selected Papers of Ludwig Edelstein.* Baltimore: Johns Hopkins Press, 1967.

Thomas, Clayton L., ed. *Taber's Cyclopedic Medical Dictionary,* 17th ed. Philadelphia: F.A. Davis, 1993.

"Thomas, E. Donnall—Autobiography." The Nobel Foundation. http://www.nobel.se/medicine/laureates/1990/thomas-autobio.html.

Thomas, E. Donnall. "Bone Marrow Transplantation—Past, Present and Future." Nobel lecture, Stockholm, Sweden, December 8, 1990.

Tucker, K. Kay. "Osteology." Northern Arizona University. http://jan.ucc.nau.edu/~kkt/EXS334/Osteology.html.

"What Is Scleroderma?" Byfield, MA: Scleroderma Foundation, 2000.

Wright, Vonda J., et al. "Muscle-Based Gene Therapy and Tissue Engineering for the Musculoskeletal System." *Drug Discovery Today* 6, no. 14 (July 2001): 728–733.

Index

About the Author

EVELYN KELLY is an independent scholar who has written 10 books and over 400 journal articles. She is a member of the National Association of Science Writers, was past president of American Medical Writer's Association, Florida chapter, and was recently invited to join the American Society of Journalists and Authors.